T0332683

Radar Scattering from Modulated Wind Waves

Radar Scattering from Modulated Wind Waves

Proceedings of the Workshop on Modulation of Short Wind Waves in the Gravity-Capillary Range by Non-Uniform Currents, held in Bergen aan Zee, The Netherlands, 24-26 May 1988

Edited by

G. J. Komen

and

W. A. Oost

Royal Netherlands Meteorological Institute (KNMI), De Bilt, The Netherlands

KLUWER ACADEMIC PUBLISHERS

DORDRECHT / BOSTON / LONDON

Library of Congress Cataloging in Publication Data

Workshop on Modulation of Short Wind Waves in the Gravity-Capillary
Range by Non-Uniform Currents (1988 : Bergen aan Zee, Netherlands)
Radar scattering from modulated wind waves : proceedings of the
Workshop on Modulation of Short Wind Waves in the Gravity-Capillary
Range by Non-Uniform Currents / edited by G.J. Komen and W.A. Oost.
p. cm.
"Bergen aan Zee, the Netherlands, 24-26 May 1988."
Includes indexes.
ISBN 0-7923-0146-3 (U.S.)
1. Ocean waves--Remote sensing--Congresses. 2. Radar--Congresses.
I. Komen, G. J. II. Oost, W. A. (Wiebe Aaldert), 1938-
III. Title.
GC206.W67 1988
551.47'022--dc19 89-2820

ISBN 0-7923-0146-3

Published by Kluwer Academic Publishers,
P.O. Box 17, 3300 AA Dordrecht, The Netherlands.

Kluwer Academic Publishers incorporates
the publishing programmes of
D. Reidel, Martinus Nijhoff, Dr W. Junk and MTP Press.

Sold and distributed in the U.S.A. and Canada
by Kluwer Academic Publishers,
101 Philip Drive, Norwell, MA 02061, U.S.A.

In all other countries, sold and distributed
by Kluwer Academic Publishers Group,
P.O. Box 322, 3300 AH Dordrecht, The Netherlands.

printed on acid free paper

Printed in The Netherlands

CONTENTS

SESSION 4
PERTURBATION OF THE GRAVITY-CAPILLARY WAVE SPECTRUM BY CURRENT VARIATION
Chairman: W. Alpers

SESSION 5
GENERAL DISCUSSION

EDITOR'S PREFACE.

Ten years ago, de Loor and co-workers at TNO, The Netherlands, were the first to report bottom topography patterns in real aperture radar (RAR) images of the southern North Sea. At that time, this was a real puzzle. The skin depth of microwaves for sea water is only of the order of centimeters while the sea bottom is about 20 meters below the surface. Electromagnetic radiation therefore cannot probe the bottom directly. Similar phenomena were found in radar imagery from SEASAT and SIR-A/B synthetic aperture radars (SAR's) of Nantucket Shoals, the English Channel and many other coastal areas. Since then theory and ocean field experiments (i.e., Phelps Bank, Georgia Straits, SARSEX, TOWARD, FASINEX, etc.) have advanced our understanding considerably. We now know that these surface signatures are the results of surface currents, perturbed by the bottom topography, which refract the propagation and modulate the energy of (short) surface waves so as to cause microwave backscatter power variations. Hence, any large scale ocean features containing nonuniform surface currents (i.e. internal waves, eddies, fronts, etc.) will cause similar manifestations in the radar imagery by means of current-wave-microwave interactions. Observations confirm this.

The whole issue of modulation of radar backscatter by oceanographic features has received increased attention because of the development of the scatterometer as a wind sensor. In fact, SEASAT not only provided SAR pictures, but also the first surface winds over the sea measured from space with a microwave scatterometer. The basic assumption behind this technique is that radiation from the scatterometer on board the spacecraft is received back from the sea surface through scattering from small sea waves with approximately the same wavelength as the microwaves (a few cm in the case of the SEASAT scatterometer). These small sea waves are generated locally by the wind, and so the backscattered radiation is a measure for the wind at the sea surface. This technique, which is going to be implemented again on the oncoming ERS-1 satellite, gives global coverage of the wind field over the oceans. It is evident that a correct interpretation of the scatterometer data requires a good understanding of the oceanographic processes responsible for short wave modulation. There is a vast variety of such processes. In addition to nonuniform currents we have modulation and tilting by the larger ones, as well as the effect of surfactants and temperature variations on wave spectra and wind-wave relations. Breaking waves and other nonlinear effects furthermore complicate the situation.

In a discussion between Dr. G.R. Valenzuela of the Naval Research Laboratory and one of us (GJK), it was felt that time was ripe for a review of recent experimental and theoretical research on the modulation of short gravity-capillary waves by non-uniform currents, and related topics, such as microwave probing and imaging mechanisms.

Such a review could best be obtained in the interactive atmosphere of a workshop with a group of invited scientists, active in all related disciplines. An organizing committee was therefore established, consisting of Dr. G.R. Valenzuela, Prof. W. Alpers of the University of Bremen (FRG), Dr. K. van Gastel (University of Utrecht, the Netherlands) and the present editors. The workshop was supported by the US Office of Naval Research and the Dutch National Remote Sensing Board. It took place on May 24-26, 1988 in the Dutch coastal resort of Bergen aan Zee under the title: "The Modulation of Short Wind Waves in the Gravity-Capillary Range by Non-Uniform Currents". The meeting was characterized by enthusiastic and lively contributions which made it an inspiring and fruitful event. It is not without pride that we present the proceedings.

G.J. Komen and W.A. Oost.

Session 1

Tidal currents and Bathymetry

Chairperson: G. P. de Loor

DETERMINING THE CURRENTS OVER PHELPS BANK

DAVID A. GREENBERG
Physical and Chemical Sciences Branch
Department of Fisheries and Oceans
Bedford Institute of Oceanography
Dartmouth Nova Scotia
Canada B2Y 4A2

GASPAR R VALENZUELA and DAVIDSON T CHEN
Space Sensing Branch
Space Systems Technology Department
Naval Research Laboratory
Washington, D. C., 20375-5000
USA

ABSTRACT. The M_2 tidal currents over Phelps Bank are examined using a fully nonlinear model. Simulations with resolutions of 2'× 2' and 0.5'× 0.5' were made of the full area covering Nantucket Shoals south of Cape Cod. A fine resolution model (0.125'× 0.125') was also run over a limited area surrounding Phelps Bank. Agreement was good in amplitude and phase when compared to surface elevation data. The agreement with current meter data was fair in amplitude and poor in phase with the exception that agreement was good at the one Phelps Bank mooring when compared with models of the full area. The limited area model showed signs of instability that were damped out with a horizontal eddy viscosity. Agreement of this model with data from the one available current meter mooring was worse. Different options are available to go from model predictions to the currents necessary for calibration of microwave data obtained over Phelps Bank.

1. Introduction

There has been a large effort to collect and examine microwave data over Phelps Bank (e.g. see Valenzuela et al., 1985). Images (Valenzuela et al., 1983) of microwave backscatter over shallow areas have exhibited features that closely mirror the bottom topography. It is felt that these images reflect the modification of short gravity waves by currents with magnitude roughly inversely proportional to the depth. For a proper calibration of the microwave signal, the currents should be accurately known.

The approach taken here is to use a barotropic numerical model to determine the M_2 tidal currents, and use these as a basis for future

G. J. Komen and W. A. Oost (eds.), Radar Scattering from Modulated Wind Waves, 3–17.

determination of the full tidal signal, which, it will be argued, constitutes the major part of current energy.

2. Do Barotropic Tides Reflect Surface Currents?

There is ample evidence that the barotropic tidal current can give a reasonable approximation to the surface currents over Phelps Bank. Garrett et al. (1978) and Loder and Greenberg (1986) have shown that in terms of the mixing parameter h/U^3 Nantucket Shoals is an area which is vertically well mixed by the turbulence associated with strong tidal currents. The current meter data collected over Phelps Bank (Greenewalt et al, 1983, table IV) show very little variation with depth, in the M_2 tidal parameters, except near the bottom. CTD data (Kaiser, 1983) confirm that the area of interest is well mixed. Moody et al. (1984) have found that in this area, over 93% of the variance in surface elevation and over 80% of the variance in currents are in the five major tidal constituents. The M_2 component is by far the dominant constituent in each case. The barotropic model of Greenberg (1979, 1983) has been used to produce an atlas of tidal currents for the Bay of Fundy and Gulf of Maine (Anon,1981). The modelled M_2 tides are shown and the mariner scales them according to the predicted tidal range at Saint John N.B. The feedback from the discriminating community of sailors who sail in the biannual Marblehead (Boston) to Halifax N.S. race , has been quite positive. Given the above, we have at least a good fighting chance of obtaining decent surface currents based on the results of the barotropic tidal model.

3. The Mechanics of the Model

The equations of continuity and motion used in the model are:

$$\frac{\partial \zeta}{\partial t} - \frac{1}{R\cos\theta}\left[\frac{\partial}{\partial \Phi}(h+\zeta)u - \frac{\partial}{\partial \theta}\cos\theta(h+\zeta)v\right] = 0 \qquad (1)$$

$$\frac{\partial u}{\partial t} - \frac{u}{R\cos\theta}\frac{\partial u}{\partial \Phi} + \frac{v}{R}\frac{\partial u}{\partial \theta} - \frac{uv\tan\theta}{R}$$

$$= v\,2\omega\sin\theta + \frac{g}{R\cos\theta}\frac{\partial \zeta}{\partial \Phi} - Ku\frac{(u^2+v^2)^{1/2} + V}{h+\zeta} +$$

$$+ \frac{A_h}{R^2\cos^2\theta}\frac{\partial^2 u}{\partial \Phi^2} + \frac{A_h}{R^2}\frac{\partial^2 u}{\partial \theta^2} - \frac{A_h\tan\theta}{R^2}\frac{\partial u}{\partial \theta} \qquad (2)$$

$$\frac{\partial v}{\partial t} - \frac{u}{R\cos\theta}\frac{\partial v}{\partial \Phi} + \frac{v}{R}\frac{\partial v}{\partial \theta} + \frac{u^2\tan\theta}{R}$$

$$= - u2\omega\sin\theta - \frac{g}{R}\frac{\partial \zeta}{\partial \theta} - Kv\frac{(u^2+v^2)^{1/2} + V}{h+\zeta} +$$

$$+ \frac{Ah}{R^2\cos^2\theta}\frac{\partial^2 v}{\partial \Phi^2} + \frac{Ah}{R^2}\frac{\partial^2 v}{\partial \theta^2} - \frac{Ah\tan\theta}{R^2}\frac{\partial v}{\partial \theta} \qquad (3)$$

Where:

ζ - sea surface elevation
t - time
R - the earth's radius
θ - latitude
Φ - longitude
u - eastward velocity
h - still water depth
ω - angular rate of the earth's rotation
g - gravitational acceleration
K - bottom friction coefficient
Ah - horizontal eddy viscosity coefficient
v - northward velocity
V - frictional background current speed

The finite difference scheme used is similar to that employed in the fine grid model described in Greenberg (1983) but adapted to spherical polar coordinates by deMargerie (1985). There were three models run in the course of this investigation: (1) a coarse resolution model with grid spacing 2'× 2' (Figure 1), (2) a medium resolution model with grid spacing 0.5'× 0.5' covering the same area, and (3) a limited area model (see Figure 1) with a resolution of 0.125'× 0.125'. The reason for looking at the finer resolution models was to try as best possible to resolve the steep topography surrounding Phelps Bank (Figure 2).

Damping is accounted for by the possible inclusion of three different parameters: the bottom friction coefficient (K = 0.0027), the mean background current speed ($V = 0.2\ m \cdot s^{-1}$) which crudely represents the damping effects of unmodelled currents (e.g. other tidal constituents and wind effects), and the horizontal eddy viscosity coefficient ($Ah = a \cdot h\ m^2 s^{-1}$). This depth dependent form of Ah is similar to that used by Schwiderski (1980). For the two models covering the full area, a = 0 was used. For the limited area, fine grid model, values of a = 0.2 and 0.4 were used to help damp out a spurious instability. This model showed poorer agreement with the available observations.

Along coastal boundaries the normal velocity is set to zero. On the open boundaries the elevation is specified. The tidal amplitudes and phases used in the specification were obtained from the Greenberg (1983) model and from the Moody et al. (1984) observations, for the limited area model these were obtained from the medium resolution model. Neither of these two sources gives adequate information in the northeast corner of the model around Martha's Vineyard. Some adjustments to the open boundary input were made to try to tune the model to get good agreement with the observations.

4. Model Results

The elevation and current amplitudes and phases were similar for the two full area models. Plots from the 2′× 2′ calculations are shown. The amplitude of the M_2 tidal elevation (Figure 3) increases from the shelf edge (40 cm) into the Gulf of Maine (60+ cm), but has a minimum (less than 20 cm) from Nantucket Island through the central part of Nantucket Shoals. The phase diagram of the M_2 tide (Figure 4) illustrates how the tidal wave approaches the area from the ocean side and the Gulf of Maine, with the latest arrival of high water being to the east of Nantucket Island. The tidal current ellipses (Figure 5) have a clockwise rotation. The (horizontal) tidal excursion is over 10 km over much of Nantucket shoals. M_2 currents of over 1 $m \cdot s^{-1}$ are found in the shallow areas.

Strong tidal currents over changing topography can generate mean currents (Loder, 1980). The mean currents generated in the model (Figure 6) tend to follow the depth contours except around the open boundaries where effects of the numerical scheme give rise to spurious velocities. The currents are weaker than those modelled in Greenberg (1983) and weaker than those found in observations. This suggests that the residual currents are not all locally generated by tidal rectification, and since no mean component is included in the open boundary specification,and no mean wind field effects included the model understandably cannot reproduce them.

Comparisons with the observations in Moody et. al. have been made for elevation and current. There were only two locations that can be compared to surface elevations in the model interior (Table I). In both places the agreement is excellent.

TABLE I

	ζ Amplitude(cm)			ζ Phase(°)		
		model			model	
Name	Obs.	2′	0.5′	Obs.	2′	0.5′
NSFE1	38.7	37.5	37.5	356	356	356
S	32.3	31.1	32.1	1	4	3

TABLE I. The observed M_2 amplitudes and phases of sea level compared with the results from the two models covering the full area. The phases are given relative to time zone 0 (UTC).

The comparisons with current measurements are less favorable (Table II- eastward velocity, Table III- northward velocity). There is fair agreement overall in amplitude, but there are widely fluctuating differences between the observed and modelled phase lags. Where there was more than one current meter on a string, some scatter is evident in the observations. Going to the finer resolution model produced changes in model results. The model lags are generally smaller than those observed, but there is so much scatter that it is difficult to call the errors systematic. An error of 29° in phase is equivalent to an error in timing of approximately one hour. Given the good agreement with the tidal elevations, which are usually more accurately measured than currents, it is hard to account for the discrepancies. Fortunately the observations of the tidal current components in the area of interest over Phelps Bank (Greenewalt et al., 1983, Table IV) are in better agreement, and leave this exercise with some credibility.

TABLE II

	u Amplitude($cm \cdot s^{-1}$)			u Phase(°)		
		model			model	
Name	Obs.	2'	0.5'	Obs.	2'	0.5'
GSC1	28.6	27.3	24.7	89	61	59
	27.3			77		
GSC2	29.1	20.9	17.7	108	76	57
	27.9			107		
B	18.0	*	*	97	*	*
Great Round	59.2	32.1	75.1	12	33	51
NSA	7.7	27.3	22.2	40	39	34
	6.4			16		
NSB	37.	25.1	35.2	20	29	2
NSD	21.5	18.6	16.6	327	339	322
NSC	45.9	22.3	15.9	32	353	318
Pollock Rip	18.6	20.1	7.7	325	277	275
NSFE1	28.1	25.0	25.0	81	60	59
	25.6			66		
Q	18.7	18.2	15.8	84	48	37
	16.3			82		
	16.1			79		
	16.0			70		
	10.0			43		
I	34.0	37.3	34.6	71	61	62
NSE	39.5	55.4	56.1	10	33	32
Nantckt LS	30.1	35.6	32.6	97	74	76

TABLE II. The observed M_2 amplitudes and phases of the eastward component of current compared with the results from the two models covering the full area. The phases are given relative to time zone 0 (UTC). (* boundary values not computed.)

TABLE III

Name	v Amplitude(cm·s^{-1}) Obs.	model 2′	model 0.5′	v Phase(°) Obs.	model 2′	model 0.5′
GSC1	59.6	58.8	58.9	40	16	17
	48.6			33		
GSC2	70.5	56.0	57.2	40	18	14
	69.3			37		
B	57.7	56.0	55.0	26	18	19
Great Round	32.0	42.4	32.7	331	309	310
NSA	58.8	78.7	74.0	344	317	315
	59.3			319		
NSB	62.9	71.5	73.5	345	302	305
NSD	41.5	50.7	46.1	345	275	270
NSC	14.9	30.3	45.0	247	265	338
Pollock Rip	66.1	40.9	30.2	307	258	64
NSFE1	24.7	18.8	19.6	353	333	331
	22.7			341		
Q	17.3	*	*	6	*	*
	14.5			4		
	14.9			3		
	13.8			351		
	7.4			314		
I	24.6	24.0	20.9	347	333	340
NSE	35.0	36.8	34.2	267	317	317
Nantckt LS	35.1	36.9	36.1	29	15	15

TABLE III. The observed M_2 amplitudes and phases of the northward component of current compared with the results from the two models covering the full area. The phases are given relative to time zone 0 (UTC). (* boundary values not computed.)

TABLE IV

PARAMETER	MEASUREMENT(at Depth)			MODEL			
	5m	13m	21m	2′	0.5′	0.125′ a=40	0.125′ a=20
MEAN(cm/s) Strength	21.8	15.4	15.8	2.8	4.9	4.0	4.4
MEAN Direction	214	218	209	180	189	181	183
MAJOR AXIS(cm/s)	66.3	64.8	62.7	69.8	73.4	56.0	57.0
MINOR AXIS(cm/s)	-44.4	-43.6	-37.7	-37.6	-38.1	-17.3	-19.3
PHASE($^{\circ}$)	29	29	29	26	32	16	19
ORIENT.($^{\circ}$)	61	63	35	50	42	61	62

TABLE IV. Comparison of the mean currents and tidal ellipse parameters obtained from the current meter mooring over Phelps Bank (Greenwalt et al., 1983) and the different model runs. The mooring was located at 45° 50.07′ N, 69° 19.81′W (see Figure 2) in a depth of 30 ± 2m.

5. From Modelled M_2 To Surface Currents

Given the problems with the limited area model, it is felt that the best surface currents should be obtained using interpolations from the 0.5′ model. These model currents then have to be adjusted in some way to account for the other tidal constituents. An attempt is also made to account for the mean current.

As mentioned earlier, the model results underestimate the strength of the mean current, most likely because there is no contribution from outside the model area or from meteorological forcing. The approach to be taken in getting this mean is to use the model produced currents, (which give us an approximate direction and our only estimate of their relative strength spatially), and scale the magnitude up to match the observations, (even though it is recognized that the mean regime may differ in pattern from the model's tidally rectified form). Since the mean current is only a fraction of the tidal signal, this method should allow for its approximate inclusion in our predictions.

There are three methods being considered for obtaining the full tidal signal. The first is to do a simple scaling based on the tide range at a nearby port, such as is done with the Atlas of Tidal Currents (Anon, 1981). There is a problem with this method in that the characteristics of the tides in the area fall between the characteristics of the two nearest appropriate ports for scaling (Boston and Newport R. I.), thus the scaling might not be appropriate to either. The second method would be to make a full tidal prediction using constituents obtained from scaling model results. Thus the modelled M_2 would be used to scale the semidiurnal constituents, and more model runs would be necessary to obtain K_1 which would be used to scale other diurnal components. The constituents thus obtained could be used in tidal predictions. The third method would be to model all five important tidal constituents for a more complete modelling produced picture, but given that it takes about 15 CPU hours on a Cray XMP to make a complete M_2 run, it seems less likely that this route will be taken.

The method of obtaining the surface currents has yet to be chosen, but it will depend on available resources.

6. Concluding Remarks

There are many possible sources of error in trying to obtain surface currents from a barotropic model of a single tidal constituent. It is hoped that the methodology described above has been justified as a reasonable first attempt.

7. Acknowledgements

The authors are very much indebted to Dr. Hsue C. Tung from Bendix Corporation for much help in the computations carries out at NRL.

8. References

Anon 1980: *Atlas of Tidal Currents - Bay of Fundy Gulf of Maine*, Canadian Hydrographic Service, Ottawa.

deMargerie, S., 1985: 'Barotropic tidal circulation of the Scotian Shelf', *Contract Rep. Can. Hydr. Ser., OSC 84-00218*, Dartmouth Canada.

Garrett, C. J. R., J. R. Keeley, and D. A. Greenberg, 1978: 'Tidal mixing versus thermal stratification in the Bay of Fundy and Gulf of Maine', *Atmosphere-Ocean*, **16**, 403-423.

Greenberg, D. A., 1979: 'A numerical model investigation of tidal phenomena in the Bay of Fundy and Gulf of Maine', *Mar. Geod.*, **2**, 161-187.

Greenberg, D. A., 1983: 'Modeling the mean barotropic circulation in the Bay of Fundy and Gulf of Maine', *J. Phys. Oceanogr.*, **13**, 886-904.

Greenewalt, D., C. Gordon, and J. McGrath, 1983: *Eulerian Current Measurements at Phelps Bank*, NRL Memo Rep. 5047, US Naval Research Laboratory, Washington, DC, 51pp.

Kaiser, J. A. C., 1983: *Data Validation and Summary for the Remote Sensing Experiment: Phelps Bank, July, 1982 Part I: Hydrography*, NRL Memo Rep. 5165, US Naval Research Laboratory, Washington, DC, 146pp.

Loder, J. W., 1980: 'Topographic rectification of tidal currents on the sides of Georges Bank', *J. Phys. Oceanogr.*, **10**, 1399-1416.

Loder, J. W.,and D. A. Greenberg, 1986: 'Predicted positions of tidal fronts in the Gulf of Maine region', *Continental Shelf Research*, **6**, 397-414.

Moody, J. A., B. Butman, R. C. Beardsley, W. S. Brown, P. Daifuku, J. D. Irish, D. A. Mayer, H. O. Mofjeld, B. Petrie, S. Ramp, P. Smith, and W. R. Wright, 1984: 'Atlas of tidal elevation and current observations on the northeast American continental Shelf and slope', *U. S. Geological Survey Bulletin 1161*, Alexandria, VA.

Schwiderski, E., 1980: 'On charting global ocean tides', *Rev. Geophys. Space Phys.*, **18**, 243-268.

Valenzuela, G. R., D. T. Chen, W. D. Garrett, and J. A. C. Kaiser, 1983: 'NRL remote sensing experiment' *EOS*, **64**, 618-619.

Valenzuela, G. R., W. J. Plant, D. L. Schuler, D. T. Chen, and W. C. Keller, 1985: 'Microwave probing of shallow water bottom topography in the Nantucket Shoals', *J. Geophys. Res.*, **90**, 4931-4942.

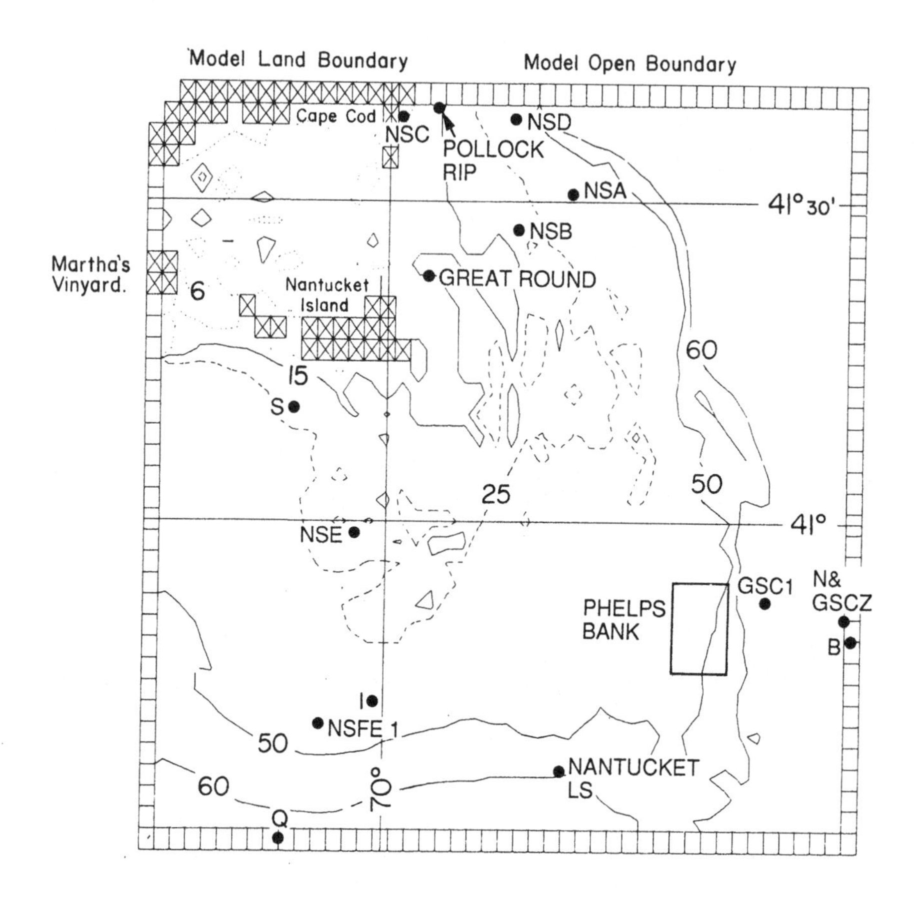

Figure 1. The model domain for the 2′(shown) and 0.5′ models. The depth contours are given in metres. The observation points which are compared to model results are indicated. The rectangle indicates the domain of the 0.125′ model around Phelps Bank.

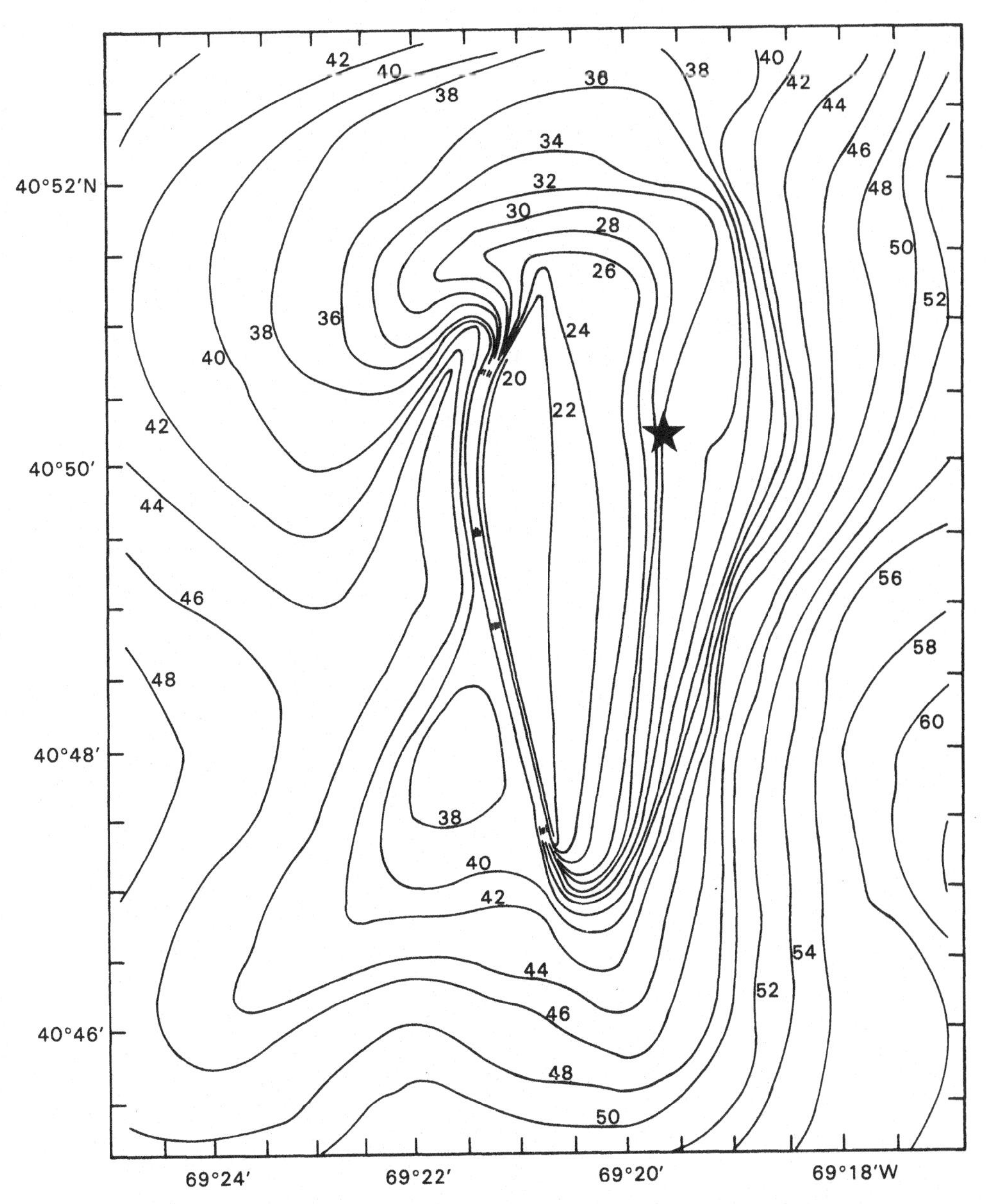

Figure 2. The detailed bathymetry of Phelps Bank. The position of the current meter mooring is indicated by the star on the northeast side of the Bank.

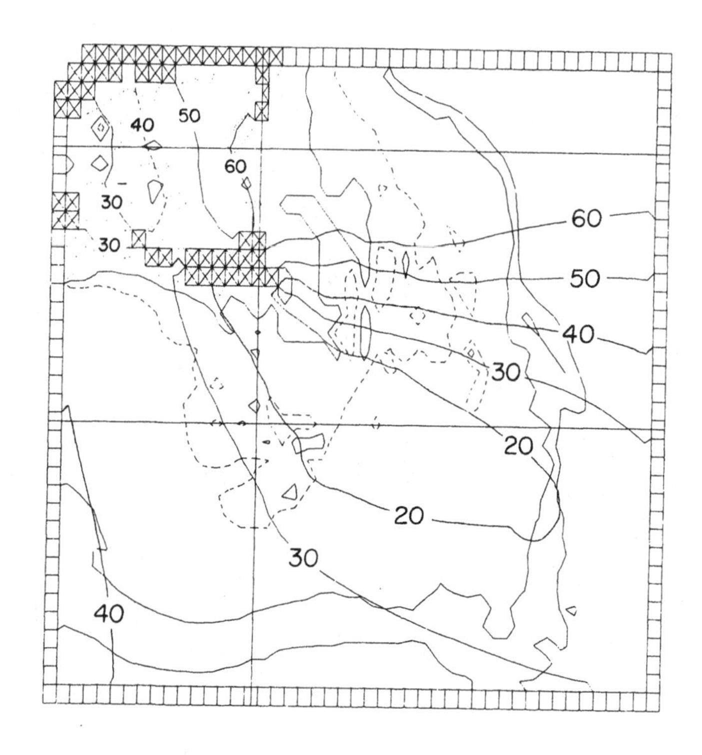

Figure 3. The lines of equal M2 tidal amplitude as produced from the 2′ model.

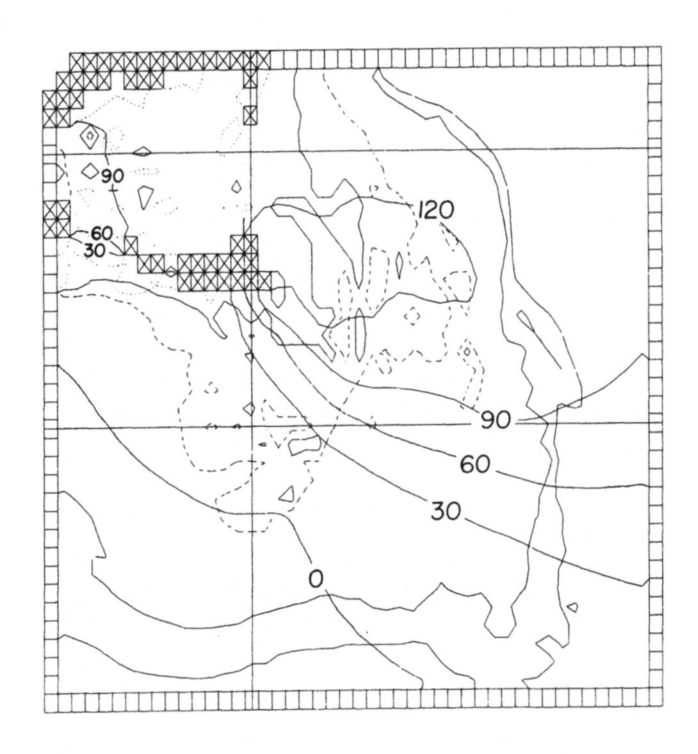

Figure 4. The co-phase lines for the M_2 tide as produced in the 2' model. Phases are relative to time zone 0 (UTC).

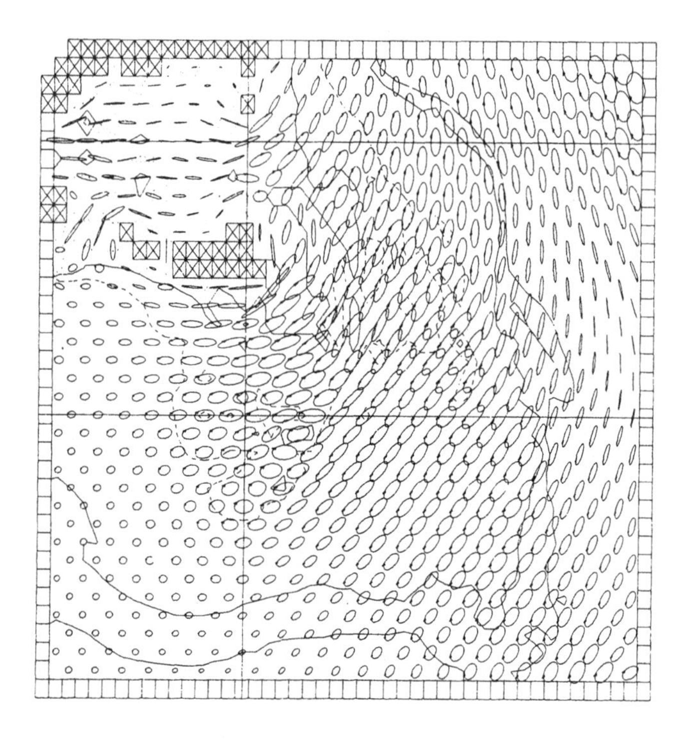

Figure 5. The M2 tidal ellipses from the 2′ model. The strength of the tidal current in a given direction is proportional to the length of the vector from the centre of the ellipse to the perimeter in that direction. The ellipses are scaled such that the path of a particle in the water would trace out an ellipse twice as large as that shown. The mark on the ellipse indicates the direction and relative strength of the current vector at the start of a tidal cycle.

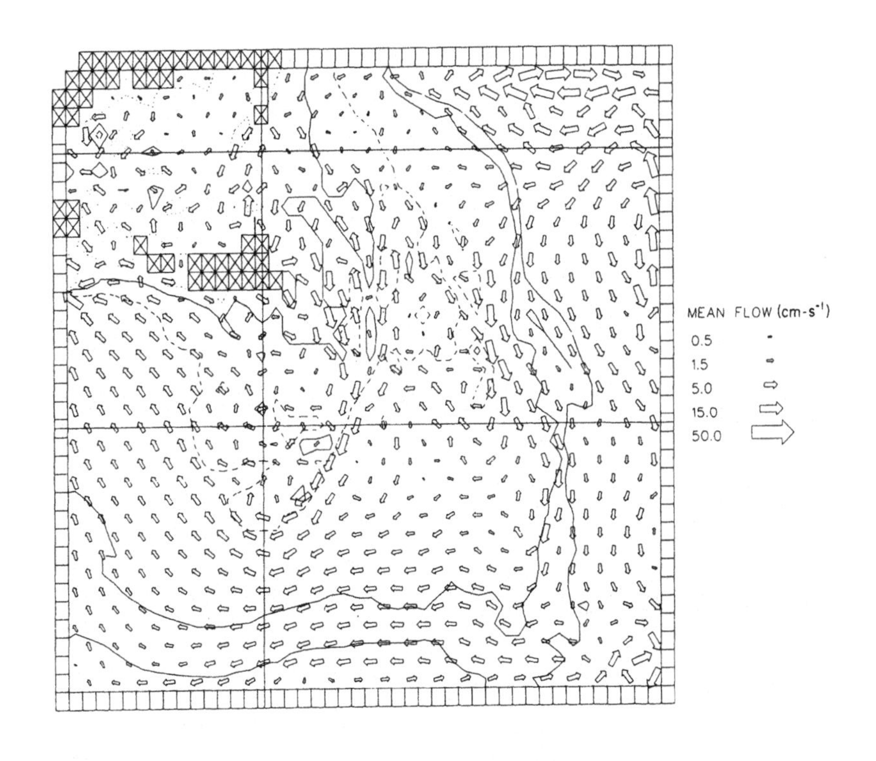

Figure 6. The mean currents from the 2′ model arising from the nonlinear terms in the model equations. The currents around the northeast boundary are the result of inadequate open boundary specification.

STUDIES OF THE SEA SURFACE WITH RADAR IN THE NETHERLANDS
A SHORT HISTORY

G.P. de Loor
Physics and Electronics Laboratory TNO
PO Box 96864
2509 JG The Hague, The Netherlands

ABSTRACT. A short survey is given of the studies of the sea with radar in the Netherlands. Topics of research are the study of waves, the radar backscatter, bathymetry, and internal waves.

1. INTRODUCTION

Due to its long shoreline and its low lie a good knowledge of the sea and its behaviour is essential for the Low Countries to survive. Therefore, with the advent of remote sensing also a major portion of the research investigating the possibilities of remote sensing has been devoted to its application at sea. All this research is executed in interdisciplinary teams of which the participants (in changing combinations) are: the Department of Water Control (RWS), the Physics and Electronics Laboratory TNO (FEL-TNO), the Delft University of Technology (DUT), the Royal Netherlands Meteorological Institute (KNMI), Delft Hydraulics and the Netherlands Aerospace Laboratory (NLR).

For groundbased measurements a research platform is available in the North Sea (Noordwijk tower). It offers the opportunity to perform experiments with radar in open sea. For experiments with airborne equipment two systems are available. The first is a digital and absolute side-looking radar. This SLAR delivers geometrically corrected images with pixel values in absolute dB's (1,2). The second system is a scatterometer (the *DUTSCAT*) which measures the radar backscatter at fixed angles (3,4) at six frequencies simultaneously.

Topics of research are the study of waves, the radar backscatter, bathymetry, and internal waves.

2. WAVES

The study of sea waves with radar has a long history in the Netherlands. Already in the late fifties *Oudshoorn* used a ships radar (35 GHz system for navigation in narrow waters) to study wave- and flow patterns in estuaries (5). With one or two wave gauges as references he was able to give the wave-height in the whole estuary together with the wave-directions. Later *Hoogeboom and Rosenthal* (6) used a similar radar to

G. J. Komen and W. A. Oost (eds.), Radar Scattering from Modulated Wind Waves, 19–24.

determine wave directional spectra at sea. Prolonged experimentation showed that it is also possible to estimate water-depth and -current in shallow water (7). A more sophisticated system is now under development at FEL-TNO (the *SHIRA* system). This system records an area of 128*128 pixels in the field of view of the radar and does so every sweep of the antenna (up to one per second). The area can be chosen at will. The wave-directional data are presented in the same form as used for the *WAVEC* buoy which is being used as a reference during the trials. Figure 1 gives an example.

A similar procedure is used with the digital SLAR system. It enables us to measure wave spectra at any place. By referencing the data to those of a *WAVEC* buoy it was possible to determine the Modulation Transfer Function M (8). The result is presented in figure 2.

3. THE RADAR BACKSCATTER

Several series of groundbased radar backscatter measurements have been made at platform Noordwijk in the course of the years and some of them in relation with international experiments as *MARSEN* and *HEXOS* (9,10). The last experiment took place in 1986 during *HEXOS*. Backscatter data were then collected in parallel with a set of other geophysical measurements taken simultaneously in the same area as illuminated by the radar and recorded simultaneously. Evaluation of the data is still in progress.

Due to the movement of the sea surface the radar reflection varies with time. Comparing this modulation of the radar backscatter with data taken simultaneously from a wave buoy the MTF was also determined. The result obtained suggests that for our data much, if not most, of the modulation of the radar backscatter can be explained by slope-modulation. This modulation is also given in figure 2 (thick line).

Radar backscatter measurements with the airborne scatterometer were collected primarily for ESA in the ESA windscatterometer campaign and the *TOSCANE* experiments (11).

All these measurements, but also those reported in the literature, show unexpected phenomena for which no good physical explanation is at hand. To cover this problem and also to develop a better understanding of the empirical relations now in use (as e.g. the windscatterometer algorithmes) and phenomena as the showing up of bathymetry and internal waves in side-looking radar imagery, it was decided to design an experiment in the large waveflume of Delft Hydraulics. This experiment, the *VIERS-1* experiment is described by *van Halsema* (12) and *Jaehne* (13).

4. BATHYMETRY

The first observations of bathymetry for the Dutch coast were made in the fall of 1969. In an experiment to detect swell with radar slowly the bottom became visible in the imagery when approaching low tide (14,15). The phenomenon was regularly observed since then and along the whole coast. It was therefore decided to do a small experiment in 1977 and to participate in the experiments around *SEASAT* in 1978, among others to acquire imagery which possibly might also show the effect.

Figure 3 gives a result of the 1977 experiment. Prolonged experimentation showed that the pattern remained stable in that area even till after one year. It was impossible at that time, however, to clearly match the features seen in the imagery with the actual bottom topography, because of the lack in the geometric accuracy of the SLAR system then in use. The only conclusion that could be drawn was that the lines in figure 3 must either be associated with the top of the underwater dunes or with areas with a large height gradient. The associated contrasts (variation in radar backscatter) proved to be of the same order of magnitude as the modulation found for seawaves.

For track 1473 of *SEASAT* (8 Oct. 1978, 00.47 GMT) imagery showing bathymetry for the Dutch coast was obtained. It fitted in general with our own SLAR data obtained earlier, but contrasts were very low, although it was low tide. This may possibly due to the fact that the imagery obtained was optically processed. No digitally processed data are yet available of the area containing figure 3.

Lack of funding prevented us from doing more work in this field. Only very recently the problem was taken up again (16).

5. INTERNAL WAVES

Internal waves were detected by us only twice and both times in very shallow water near river outflows (depth to max. 25 m). The waves seen were very short (50 m max.) and this possibly may be one of the reasons that the phenomenon was not seen before by us. The associated contrasts are very low: again in the same order as that of the modulation by seawaves. To observe such a phenomenon the windspeed may not be too large: 8 - 10 m/s max., otherwise the sea becomes too choppy. This means a low backscatter for the system now in use by us: high resolution (7.5 m) and HH polarization. The older system with which the bathymetry was studied had a poor resolution (30 m) - which is sufficient for that application - but therefore had a higher sensitivity and a higher echo per pixel.

REFERENCES

1. Hoogeboom, P., 1983, 'Preprocessing of side-looking airborne data'; *Int.J.Remote Sensing*, **4**, 631-637
2. Hoogeboom, P, P. Binnenkade and L.M.M. Veugen, 1984, 'An algorithme for radiometric and geometric correction of digital SLAR data'; *IEEE Trans. Geosci and RS*, **GE-22**, 570-576
3. Snoey, P., 1985, 'Analysis and calibration of the Dutch scatterometer measurements'; *Procs 3rd Int. Colloquium on Spectral Signatures of Objects in Remote Sensing'*, Les Arcs, France, 16-20 Dec., ESA publication SP-247, 49-51
4. Snoey, P. and P. Swart, 1987, 'The DUT airborne scatterometer'; *Int.J.Remote Sensing*, **8**, 1709-1716
5. Oudshoorn, H.M., 1960, 'The use of radar in hydrodynamic surveying', Chapter 4 in: J.W. Johnson (editor), *Procs. of*

the seventh conference on coastal engineering', The Hague, Aug.

6. Hoogeboom, P. and W. Rosenthal, 1982, 'Directional wavespectra in radar images', Paper WA-2 in: *Procs IGARSS'82 Symposium*, Munich, June 1-4; IEEE Catalog No. 82CH14723-6
7. Hoogeboom, P., J.C.M. Kleyweg and D. van Halsema, 1986, 'Seawave measurements using a ships radar'; *Procs IGARSS'86 Symposium*, Zurich, 8-11 Sept.; ESA Publication SP-254
8. Hoogeboom, P., H.C. Peters and H. Pouwels, 1988, 'Comparison of wave parameters determined from SLAR images and a pitch and roll buoy'; *Int.J.Remote Sensing*, **9**, in print
9. De Loor, G.P. and P. Hoogeboom, 1982, 'Radar backscatter measurements from platform Noordwijk in the North Sea'; *IEEE J. Oceanic Eng.*, **OE-7**, 15-20
10. De Loor, G.P., 1983, 'Tower-mounted radar backscatter measurements in the North Sea, *J.Geophys.Res.* **88**, No.C14, Nov.20, 9785-9791
11. Group of authors, 1985, 'The PROMESS and TOSCANE T simulation campaign', *Procs. 3rd International Colloquium on Spectral Signatures of Objects in Remote Sensing'*, Les Arcs, France, 16-20 Dec., ESA Publication SP-247, 3-82
12. Van Halsema, D., 1988, 'First results of the VIERS-1 experiment', this workshop
13. Jaehne, B., 1988, 'Energy balance in small scale waves', this workshop
14. De Loor, G.P. and H.W. Brunswijk van Hulten, 1978, 'Microwave measurements over the North Sea', *Boundary-Layer Meteorology*, **13**, 119-131
15. De Loor, G.P., 1981, 'The observation of tidal patterns, currents and bathymetry with SLAR imagery of the sea', *IEEE J. Oceanic Eng.*, **OE-6**, 124-129
16. Vogelzang, J., 1988, 'The mapping of underwater bottom topography with SLAR', this workshop

FIGURES

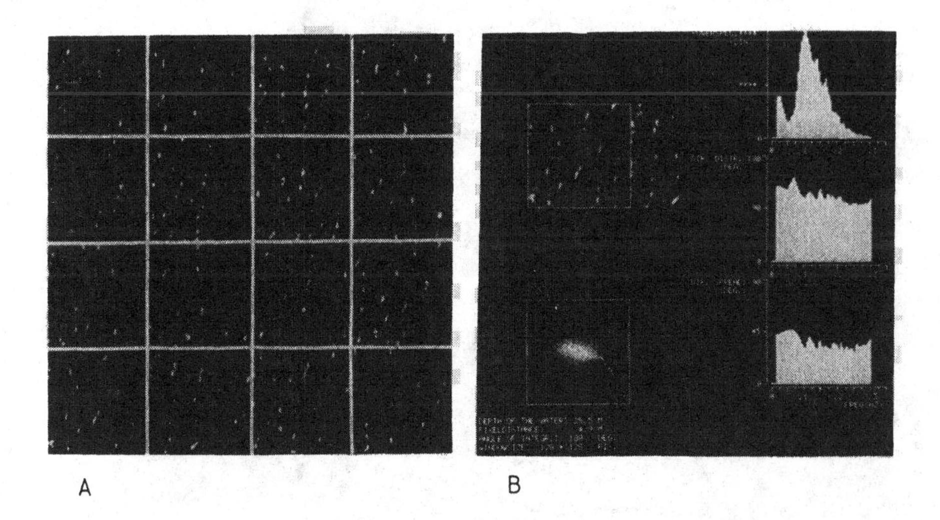

Figure 1. The measurement of wave-directional spectra with the *SHIRA* equipment. A. 16 successive images as recorded by the equipment (16 seconds, 128*128 pixels); B. directional wave spectrum as determined from one recording (image 13).

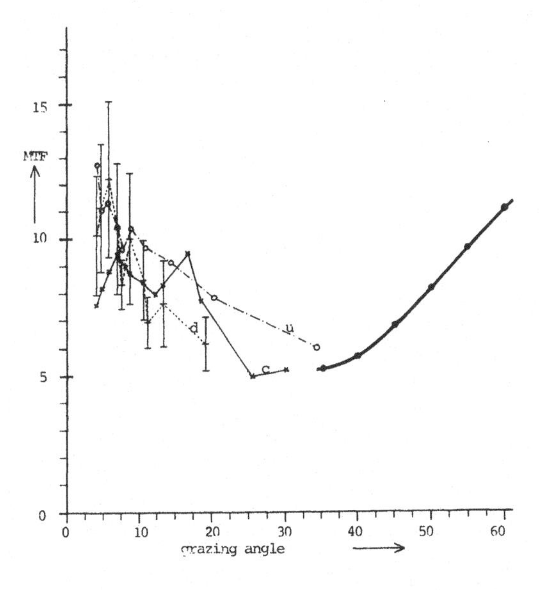

Figure 2. The Modulation Transfer Function M as found for three look directions (u: upwind, d: downwind, c: crosswind) with the Dutch digital SLAR compared with the MTF as found from measuremants with scatterometers (thick line).

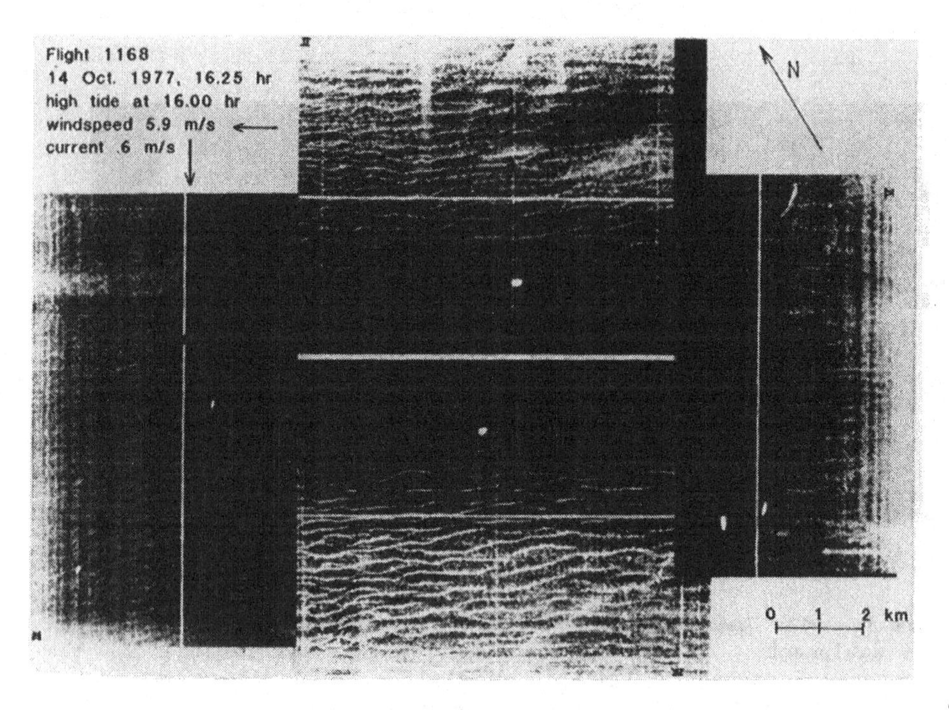

Figure 3. Four radar recordings of the bottom 20 km for the Dutch coast near Noordwijk taken near high tide and from different directions.

Session 2

Electromagnetic scattering

Chairperson: G. J. Komen

CALCULATION OF MICROWAVE DOPPLER SPECTRA FROM THE OCEAN SURFACE WITH A TIME-DEPENDENT COMPOSITE MODEL

D.R. Thompson
Johns Hopkins University
Applied Physics Laboratory
Laurel, Maryland 20707
U.S.A.

ABSTRACT. Doppler spectra are computed using a time-dependent composite model. This model reduces to the proper specular and small-perturbation limits for V-V and H-H polarization, and its time dependence is based on the assumption of linear surface evolution. The width of the computed spectra increases with increasing radar frequency and decreasing incidence angle due primarily to the motion of the long-wave surface. This motion also produces a shift in the spectral peak away from the Bragg frequency. This shift, which is largest at lower incidence angles and higher radar frequencies, is also due to the effect of long-wave motion on the backscattered field. At incidence angles ≥ 45 deg, where most of the Doppler measurements exist, the calculations are in good general agreement with the data.

1 Introduction

Because of the rapid advances in ocean remote sensing with microwave radars during the past several years, there is renewed interest in the physics which governs the scattering of microwave radiation from the ocean surface. A knowledge of how the evolution of the ocean surface-wave spectrum affects the scattered electromagnetic field, for example, is essential for a quantitative understanding of the properties of measured microwave Doppler spectra. It is clear that remote sensing will not provide a reliable alternative to in-situ measurements of ocean surface until the connection between the measured radar return and the behavior of the surface is understood in detail.

Such an understanding is made more difficult because of the complicated hydrodynamics which is sometimes needed to describe the motion of ocean surface waves. One can however make "reasonable" assumptions about the surface behavior and then use these assumptions in the scattering models to calculate the scattered fields and compare with measurements. As will be shown later in this paper, it is possible to describe seemingly complicated features of the scattered fields with relatively simple assumptions about the surface structure.

The primary goal of our study is to present calculations of Doppler spectra for microwave radiation scattered from the ocean surface as a function of wind conditions, radar frequency, incidence angle, and polarization. These calculations are based on a composite-type model which is a generalization of the model derived from the

G. J. Komen and W. A. Oost (eds.), Radar Scattering from Modulated Wind Waves, 27–40.

Kirchhoff approximation and discussed in [1]. The present model is based on an expression for the backscattered field given by Holliday [2] which is derived using two iterations of the surface-current integral equation, and which reduces to the field given by the small perturbation method in the limit of small amplitude surface roughness. In order to compute Doppler spectra, we have further modified the results of [1] to include the effects of surface-wave propagation. We describe our scattering model in more detail in Section 2 and show some sample calculations and comparisons with data in Section 3.

2 Scattering Model

Computations of the electromagnetic field scattered from a rough surface are usually formulated in terms of the Stratton-Chu equation [3] which gives the field at an arbitrary field point in terms of an integral of the surface current over the bounding surface. The problem then reduces to one of finding the surface current. This can be done rather easily for a planar surface or when the surface roughness height is small compared to the radiation wavelength. For the latter case, the backscattered field is proportional to the component of the surface height spectral density at the Bragg wavenumber (defined as twice the projection of the radar wavenumber on the horizontal surface). For surfaces such as the ocean's, which contain a broad range of roughness scales, the problem becomes much more complicated.

One method for attacking the scattering problem is to let the field point in the Stratton-Chu equation approach the surface. This procedure yields an integral equation for the surface current which can, in principle, be solved by iteration.[1] The first iteration of this surface-current integral equation yields the Kirchhoff approximation for the scattered field [2,4] which is exact for a planar surface. The Kirchhoff approximation can also be used to predict the scattered field from more complicated surfaces. In fact it has been found [5] that intensity modulations in the radar cross section produced by internal waves can be adequately explained using this approach. Furthermore, it has been shown [1] that if the surface is partitioned into long- and short-wave components, the Kirchhoff approximation contains the essential features of a two-scale composite model.

In the small wave-height limit however, the Kirchhoff approximation does not yield the proper polarization dependence. Holliday has shown [2] that there are terms resulting from the second iteration of the surface-current integral equation of the same order (linear in the surface slope) as those occurring in the first iteration, and he has derived an expression from the first two iterations which is exact through terms linear in the surface slope. This expression gives the proper polarization dependence in the small wave-height limit.

In the present paper, we apply a procedure similar to that given in [1] to derive a composite-model expression for the scattered field based on two iterations of the

[1] For all computations to be presented in this paper, we assume that the ocean is a perfect conductor.

surface-current integral equation as discussed in [2]. This expression is given by

$$\begin{aligned}\vec{B}(\vec{r}_0,t) &= \frac{1}{2\pi i}\frac{e^{i\kappa r_0}}{r_0}\int G(\vec{x})\exp[-2i\vec{\kappa}_H\cdot\vec{x}-2i\kappa_z\eta_L(\vec{x},t)] \\ &\quad\times\Big\{[\kappa_z-\vec{\kappa}_H\cdot\vec{\nabla}\eta_L(\vec{x},t)][1-2i\kappa_z\eta_S(\vec{x},t)]\vec{B}_0 \\ &\quad+2\vec{B}_0\cdot[\hat{e}_z\times\vec{\nabla}\eta_S(\vec{x},t)]\frac{\vec{\kappa}_H\times\vec{\kappa}}{\kappa_z}\Big\}\,d\vec{x}. \end{aligned} \tag{1}$$

In Eq.(1), $\vec{\kappa}_H$ and κ_z are the horizontal and vertical components of of the radar wavevector $\vec{\kappa}$, respectively, $\vec{B}_0$ specifies the incident field strength and polarization, $G(\vec{x})$ describes the antenna footprint, $\hat{e}_z$ is a unit vector along the (vertical) z-axis, and r_0 is the distance from the center of the footprint to the antenna. Also, in deriving this expression we have partitioned the surface height $\eta(\vec{x},t)$ at position $\vec{x}$ and time t into long-scales specified by $\eta_L(\vec{x},t)$ and short-scales specified by $\eta_S(\vec{x},t)$ such that

$$\eta(\vec{x},t)=\eta_L(\vec{x},t)+\eta_S(\vec{x},t), \tag{2}$$

with $2\kappa_z\eta_S(\vec{x},t)\ll 1$. One should note that when no long waves are present ($\eta_L(\vec{x},t)=0$ for all $\vec{x}$ and t), Eq.(1) reduces to the small perturbation method limit [2,6,8]. Similarly, when no short waves are present ($\eta_S(\vec{x},t)=0$ for all $\vec{x}$ and t) it yields "specular" scattering from a long-wave surface [6,7,8]. Thus, Eq.(1) yields the proper expression for the scattered fields in both the long- and short-wave limits.

For a deterministic surface, one may compute the backscattered field directly from Eq.(1). For most cases however, the surface cannot be specified completely and one must resort to a statistical description. We assume in this work that $\eta_L(\vec{x},t)$ and $\eta_S(\vec{x},t)$ are zero-mean Gaussian random processes and that $\langle\eta_L(\vec{x},t)\eta_S(\vec{x},t)\rangle=0$. These assumptions do not necessarily mean that the scattered field, $\vec{B}(\vec{r},t)$, is also Gaussian since $\eta_L(\vec{x},t)$ appears nonlinearly in Eq.(1). If we are dealing with a surface for which $\eta_L(\vec{x},t)=0$ (that is a surface with small-scale roughness only) then $\vec{B}$ is linear in $\eta_S(\vec{x},t)$ and hence also Gaussian. Even when $\eta_L(\vec{x},t)$ is not small, the illuminated area on the surface may contain enough independent scattering centers so that the received field is still nearly Gaussian due to the central-limit theorem. In general, we expect the backscattered field to become more non-Gaussian as the mean-squared surface height increases and as the antenna footprint decreases.

With the above assumptions, we may now write the autocovariance, $R(t)$, of the backscattered field as

$$R(t)=\frac{4\pi r_0^2}{|\vec{B}_0|^2 A_{eff}}\langle\vec{B}^*(0,r_0)\vec{B}(t,r_0)\rangle, \tag{3}$$

where A_{eff} is the effective area of the footprint, and we have normalized $R(t)$ so that $R(0)$ is the mean cross section per unit area. It can be seen from Eq.(1) that, due to the assumed independence of $\eta_L(\vec{x},t)$ and $\eta_S(\vec{x},t)$, $\langle\vec{B}^*\vec{B}\rangle$ has the form of a Fourier transform of the product of a function which depends on $\eta_L(\vec{x},t)$ times a function which depends on $\eta_S(\vec{x},t)$. We may therefore use the convolution theorem to rewrite Eq.(3) in the form

$$R(t) = R_{SP}(t) + R_{TB}(t), \tag{4}$$

where

$$R_{SP}(t) = \frac{1}{\pi}\mathcal{G}_2(2\vec{\kappa}_H, \kappa_z, t), \tag{5}$$

and

$$\begin{aligned} R_{TB}(t) \quad = \quad & \frac{2}{\pi}\int \Big\{ \mid \vec{B}_0 \mid^2 \kappa_z^2 \mathcal{G}_2(\vec{k} - 2\vec{\kappa}_H, \kappa_z, t) \\ & + 2\kappa_z \mid \vec{\kappa}_H \mid (\varepsilon_y \cdot \vec{B}_0)[\vec{B}_0 \cdot (\hat{\varepsilon}_z \times \vec{k})]\mathcal{G}_1(\vec{k} - 2\vec{\kappa}_H, \kappa_z, t) \\ & + \kappa_H^2 [\vec{B}_0 \cdot (\hat{\varepsilon}_z \times \vec{k})]^2 \mathcal{G}_0(\vec{k} - 2\vec{\kappa}_H, \kappa_z, t) \Big\} \\ & \times \left[\psi_S(\vec{k}) e^{-i\omega t} + \psi_S(-\vec{k}) e^{i\omega t} \right] d\vec{k}, \end{aligned} \tag{6}$$

and we have assumed that $\vec{\kappa}$ lies in the x-z plane. The short-wave spectral density, $\psi_S(\vec{k})$, which appears in Eq.(6), is related to the autocovariance of the short-wave portion of the surface through

$$\langle \eta_S(0,t)\eta_S(\vec{x},t)\rangle = \int \psi_S(\vec{k})\cos(\vec{k}\cdot\vec{x} - \omega t)d\vec{k}. \tag{7}$$

A similar expression

$$\langle \eta_L(0,t)\eta_L(\vec{x},t)\rangle = \int \psi_L(\vec{k})\cos(\vec{k}\cdot\vec{x} - \omega t)d\vec{k}, \tag{8}$$

applies for the long-wave surface. Note that these expressions along with Eq.(2) and our assumption that $\eta_L(\vec{x},t)$ and $\eta_S(\vec{x},t)$ are statistically independent imply that the total wave-height spectrum, $\psi([\vec{k})$, is given by

$$\psi(\vec{k}) = \psi_L(\vec{k}) + \psi_S(\vec{k}). \tag{9}$$

The influence of the long-wave portion of the surface on the scattering is contained in the functions $\mathcal{G}_0, \mathcal{G}_1$, and $\mathcal{G}_2$ in Eqs.(5-6). These functions are derived from a stationary phase approximation similar to that discussed in [1]. (The details of this derivation will be given elsewhere.) They are related to the various second moments of $\psi_L(\vec{x},t)$ and are given explicitly by

$$\mathcal{G}_0(\vec{k},\kappa_z,t) = \int H(\vec{x})\exp(-i\vec{k}\cdot\vec{x}) \times exp\left\{-2\kappa_z^2[\overline{S_x^2}x^2 + \overline{S_y^2}y^2 + \overline{V^2}t^2 + 2\overline{S_{xy}}xy - 2(\alpha x + \beta y)t]\right\} d\vec{x}, \tag{10}$$

$$\mathcal{G}_1(\vec{k},\kappa_z,t) = \left(\kappa_z + \frac{\vec{\kappa}_H\cdot\vec{k}}{2\kappa_z}\right)\mathcal{G}_0(\vec{k},\kappa_z,t), \tag{11}$$

$$\mathcal{G}_2(\vec{k},\kappa_z,t) = \left[\kappa_z^2 + \vec{\kappa}_H\cdot\vec{k} + \frac{(\vec{\kappa}_H\cdot\vec{k})^2}{4\kappa_z^2}\right]\mathcal{G}_0(\vec{k},\kappa_z,t). \tag{12}$$

The function $H(\vec{x})$ in Eq.(10) is related to the footprint area and the moments of ψ_L in the exponential term in $\mathcal{G}_0$ are given by

$$\overline{S_x^2} = \int \psi_L(\vec{k})k_x^2 d\vec{k}, \tag{13a}$$

$$\overline{S_y^2} = \int \psi_L(\vec{k})k_y^2 d\vec{k}, \tag{13b}$$

$$\overline{S_{xy}} = \int \psi_L(\vec{k})k_x k_y d\vec{k}, \tag{13c}$$

$$\overline{V^2} = \int \psi_L(\vec{k})\omega^2 d\vec{k}, \tag{13d}$$

$$\alpha = \int \psi_L(\vec{k})k_x\omega d\vec{k}, \tag{13e}$$

$$\beta = \int \psi_L(\vec{k})k_y\omega d\vec{k}. \tag{13f}$$

The above equations specify the scattering process in terms of long- and short-wave portions of the surface-wave spectral density. As in all composite-type scattering models, we must choose a cutoff wavenumber which separates the spectrum. It was found in [1] that if this cutoff is chosen to be $\approx \frac{1}{3}$ the Bragg wavenumber, the Kirchhoff and composite model yield essentially identical cross sections. We therefore choose the same separation wavenumber for all the computations to be presented in the present work. Since this separation wavenumber is frequency dependent, the moments defined by Eqs.(13) will, in general, be functions of the radar frequency. Note that when only long waves are present, $R(t) = R_{SP}(t)$ and $R(0)$ reduces to the mean cross section for specular scattering [7]. When only short waves are present, $R(t) = R_{TB}(t)$ and $R(0)$ reduces to the proper (polarization-dependent) small perturbation cross section [8].

Finally, we should mention that the time dependence in our expression for $R(t)$ is based on the assumption that each component of the ocean surface-wave spectrum evolves in time according to the free-wave dispersion relation. This assumption, which is related to the assumption of Gaussian surface displacements, implies that $\omega(\vec{k})$ in Eqs.(7-8) is given by

$$\omega(\vec{k}) = \{gk[1 + (\frac{k}{k_0})^2]\}^{\frac{1}{2}} + \vec{k} \cdot \vec{U}, \tag{14}$$

where g is the acceleration of gravity, $k_0 = 363$ rad/m, and $\vec{U}$ is any background current which may be present. It is not clear at present how sensitive the backscatter computations are to the assumption of a Gaussian surface displacement. One would certainly expect that for very high seas when the presence of crested waves, breaking waves, foam, and spray becomes important, this assumption will break down. Even for more moderate sea conditions, parasitic capillary wave generation through wave-wave interactions or the modulation of short gravity/capillary waves by the orbital motion of the longer waves will introduce nonlinear effects which are very difficult to include in current models. On the other hand, the recent success of these models in describing the results of various ocean imaging experiments [1,5,9] gives us confidence that the Gaussian assumption may not be too bad.

We present in the next section some sample calculations using the time-dependent composite model discussed above, and show some comparisons with the available data. It is convenient to show these results by means of the Doppler spectrum which is simply the Fourier transform of $R(t)$ and given explicitly by

$$S(\omega) = \int_{-\infty}^{\infty} e^{-i\omega t} R(t) dt. \tag{15}$$

We will see that the Doppler spectra can sometimes exhibit rather intriguing features.

3 Doppler Spectra

In order to compute Doppler using the scattering model described in the previous section, we must specify the surface-wave spectral density $\psi(\vec{k})$. The nature of the ocean surface-wave spectrum is a topic of current research, and many questions concerning the exact form of $\psi(\vec{k})$, especially its angular dependence, remain to be answered. For the computations to be presented in what follows, we specify $\psi(\vec{k})$ in the form

$$\psi(\vec{k}) = S(|\,\vec{k}\,|) \cos^4[(\phi_k - \phi_W)/2], \tag{16}$$

where ϕ_k is the polar angle of $\vec{k}$, ϕ_W is the polar angle of the wind velocity, and $S(|\,\vec{k}\,|)$ is the Bjerkaas-Riedel spectrum which depends on the wind speed [10]. An equilibrium spectrum of this form has been used to provide a good description of the cross section modulations due to internal waves [1,9]. More flexible and detailed parameterizations of $\psi(\vec{k})$ are available - for example that of Donelan and Pierson [11]. We feel however that the features of the Doppler spectra to be presented in what follows are not too sensitive to the details of $\psi(\vec{k})$.

Before discussing Doppler spectra, we present in Figure 1 plots of the computed L-band ($\lambda_R = 0.235$ m) and X-band ($\lambda_R = 0.033$ m) cross section per unit area as a

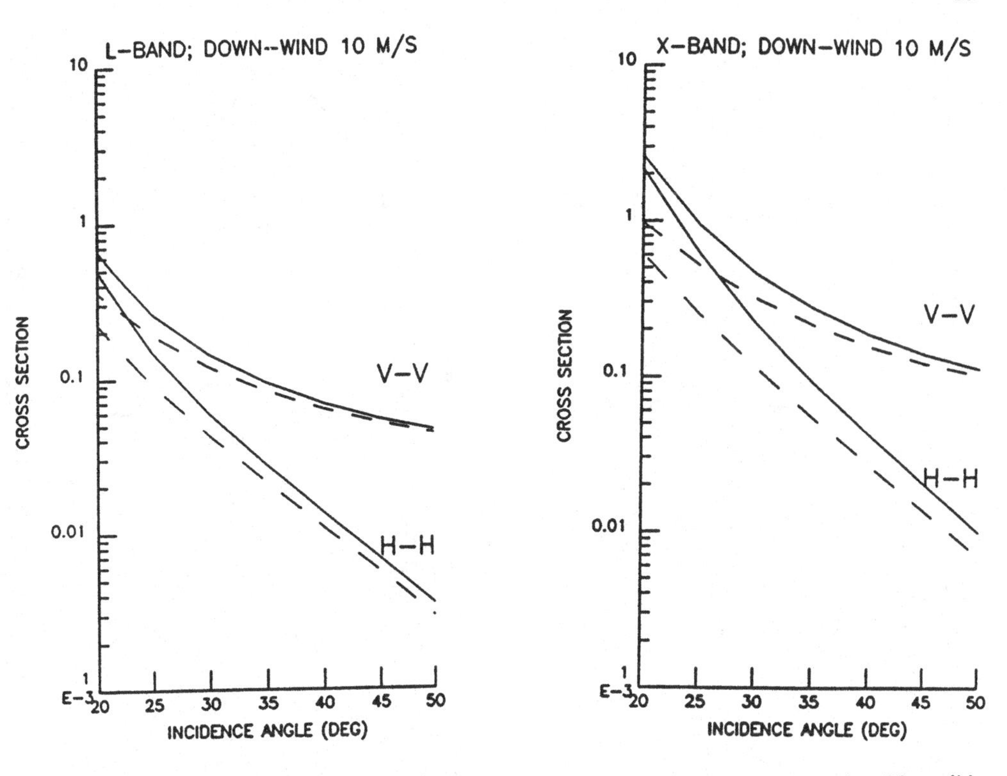

Figure 1: Plots of the L- and X-band cross section as a function of incidence angle. The solid curves show the composite-model cross sections while the dashed curves show the corresponding small-perturbation-method results.

function of incidence angle θ_I for V-V (upper solid curve) and H-H (lower solid curve) polarization. For these calculations, the wind velocity is 10 m/s along the radar look direction.[1] The dashed curves in the plots in this figure show the corresponding small-perturbation cross sections for comparison. It can be seen from Figure 1 that the cross section for H-H polarization falls off more rapidly with incidence angle than that for V-V polarization. Furthermore we see that as θ_I increases, the V-V polarization cross section becomes less and less sensitive to the long-wave surface until at $\theta_I = 45$ deg or so, the composite-model and small-perturbation cross sections are nearly equal for both L- and X-band. For H-H polarization on the other hand, a difference between the composite model and the small-perturbation method exists even at $\theta_I = 50$ deg. This result that the cross section for H-H polarization is more sensitive to the presence of long (compared to the radar wavelength) surface waves than is that for V-V polarization is an important consideration for many remote-sensing applications such as wind scatterometry [12] or the the imaging of surface

[1] In all computations in this work, we assume a $5m \times 5m$ Gaussian antenna footprint. The computations are insensitive to the footprint area as long as it is large compared to the decorrelation scale of the surface which is typically 3 or 4 Bragg wavelengths.

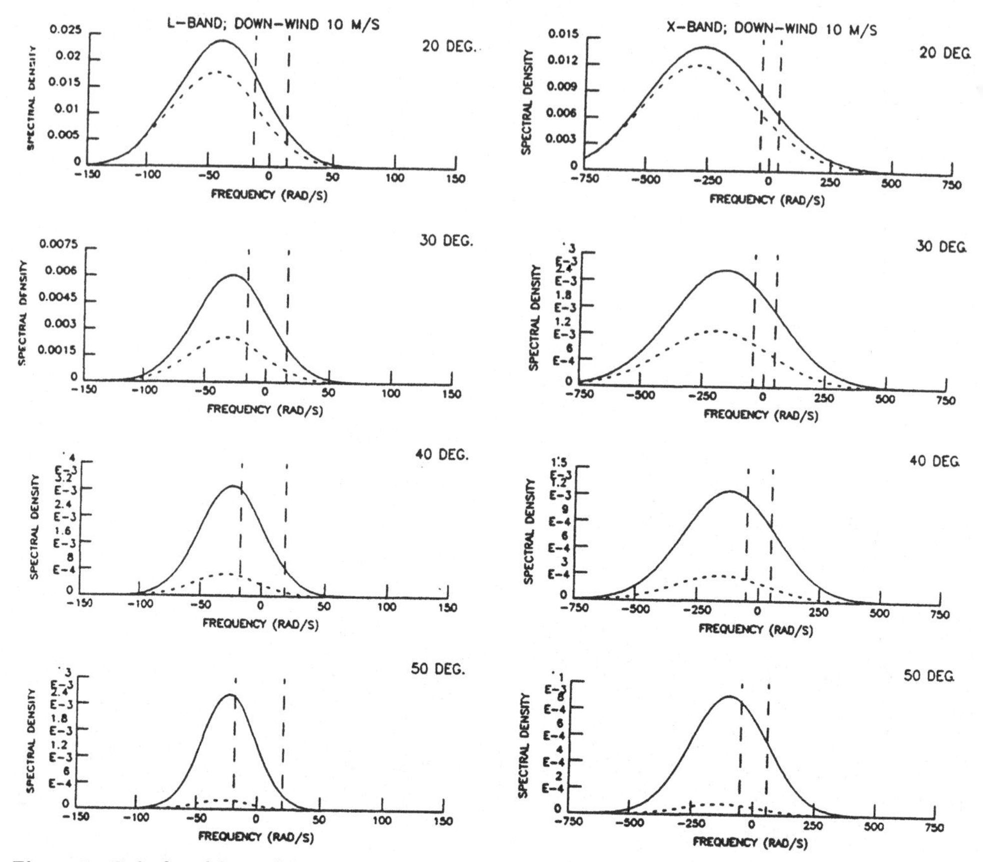

Figure 2: Calculated L- and X-band Doppler spectra for incidence angles from 20 to 50 deg and a 10 m/s wind blowing along the radar look direction. The vertical lines in the Doppler plots denote the surface-wave Bragg frequencies.

current features [1,5]. For incidence angles less than about 25 deg, the L- and X-band cross sections for both polarization states begin to deviate significantly from the small-perturbation values, and become dominated by (polarization-independent) specular scattering.

We turn now to a discussion of the computed Doppler spectra. In Figure 2, we have plotted the Doppler spectra over a range of incidence angles from 20 – 50 deg for the same radar parameters and wind conditions as in Figure 1. The dotted vertical lines in each of the plots in Figure 2 show the frequency of surface Bragg waves $(= \sqrt{gk_B}[1+(\frac{k_B}{k_0})^2]^{\frac{1}{2}}$, where the Bragg wavenumber $k_B = \frac{4\pi}{\lambda_R}\sin\theta_I)$ advancing toward (positive frequency) and receding from (negative frequency) the radar. Since the radar is looking downwind, each of the Doppler spectra is peaked at a negative frequency. Note that the L-band spectra have widths ranging from about 50 to 100 rad/s, while at X-band the widths range from about 500 to 750 rad/s or so. These widths reflect the decorrelation times of the backscattered field for the various conditions shown, and are longer for L-band than for X-band. The width of the Doppler spectra is determined primarily by the mean-squared velocity of the long waves, $\overline{V^2}$, defined in Eq.(13d). This quantity is larger at X-band since the integration over the long-wave spectrum for this case includes larger k-values than at L-band. Note that the spectral widths are nearly independent of polarization at both frequencies.

An interesting aspect of the Doppler spectra shown in Figure 2 is that the spectral peaks do not occur at the Bragg frequency as one might expect. The deviation of the peak from the Bragg frequency is most pronounced at the smaller incidence angles, and is larger at X-band than at L-band. Since we have assumed linear surface evolution in our formulation and have not included any background currents [$\vec{U} = 0$ in Eq.(14)], the observed shift is not due to advective effects. Rather, it is due to the vertical velocity or "heaving" of the long-wave surface which imparts an additional velocity along the radar line-of-sight. If one considers for a moment a surface composed of a single long wave, the origin of this shift may be understood qualitatively as follows: The vertical velocity of the long-wave surface has zero mean, but the backscatter cross section is greater for that portion of the wave whose slope is directed toward the radar. The velocity of this portion of the wave will therefore play the dominant role in producing the spectral shift. For the case presented in Figure 2 where the radar is looking downwind, the largest cross section comes from the rear face of the long wave. The component of the surface velocity along the radar line-of-sight in this region is negative (directed away from the radar), so that the observed shift in the Doppler spectra is toward more negative frequencies. If we were to repeat the calculations shown in Figure 2 with the radar look direction (or the wind direction) reversed, we would find the peak in the spectra at the corresponding positive frequency in each case. This is because for an upwind-looking radar, the largest cross section comes from the forward face of the long wave which produces a positive line-of-sight velocity in the manner just explained. The spectral peaks move closer to the Bragg frequency as the incidence angle increases since the line-of-sight component of the surface motion is proportional to $\cos\theta_I$. The spectral shifts are determined by the quantities α and β defined by Eqs.(13e-13f), respectively, which are larger for X-band than for L-band. Finally, it can be seen from Figure 2 that the spectral peak for H-H polarization is slightly more negative than that for V-V. This effect is also most pronounced for smaller incidence angles.

In Figure 3, we show Doppler spectra computed for a 2.5 m/s wind blowing perpendicular to the radar look direction. For this light wind condition, the dominant surface wave is on the order of 10 m rather than $\approx$ 100 m for the 10 m/s wind speed for the example in Figure 2. Therefore the effect of the long waves on the spectra shown in Figure 3 is significantly reduced. This can be seen quite easily by the smaller spectral widths. Since the radar is directed perpendicular to the wind direction, there are as many surface waves propagating toward the radar as away from it, and all the Doppler spectra in Figure 3 are symmetric about zero. Furthermore, due to the reduced broadening of the spectra by the long-wave motion, both the positive and negative frequency Bragg peaks are resolved for all incidence angles at L-band and

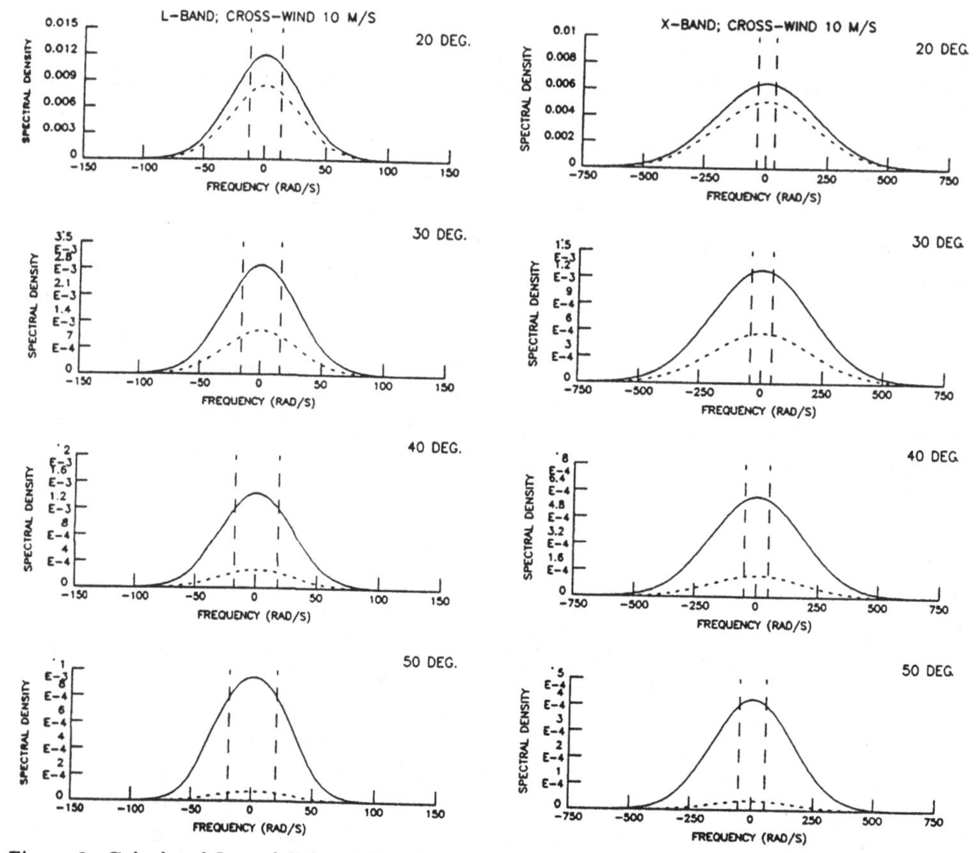

Figure 3: Calculated L- and X-band Doppler spectra for incidence angles from 20 to 50 deg and a 2.5 m/s wind blowing perpendicular to the radar look direction. The vertical lines in the Doppler plots denote the surface-wave Bragg frequencies.

for 50 deg at X-band. For X-band 40 deg incidence, the two Bragg peaks are just visible, but for lower incidence angles even with only a 2.5 m/s wind, the long-wave broadening is severe enough to obscure the Bragg peaks. As was the case for 10 m/s winds, we see from Figure 3 that there is not much difference between the V-V and H-H polarization spectra - except of course the smaller magnitude - for all the cases shown.

In order to substantiate the predictions of the scattering model presented above, we should ideally have a full knowledge of the 2-D surface spectrum $\psi(\vec{k})$ in the region covered by the antenna footprint. Measurements of the 2-D spectrum are, of course, very difficult to obtain. In lieu of these measurements, we must use empirical representations for $\psi(\vec{k})$ as discussed earlier as inputs to our models , and compare their predictions with measured Doppler spectra. However, almost all the available lower incidence angle (≤ 45 deg) backscatter measurements have been collected from

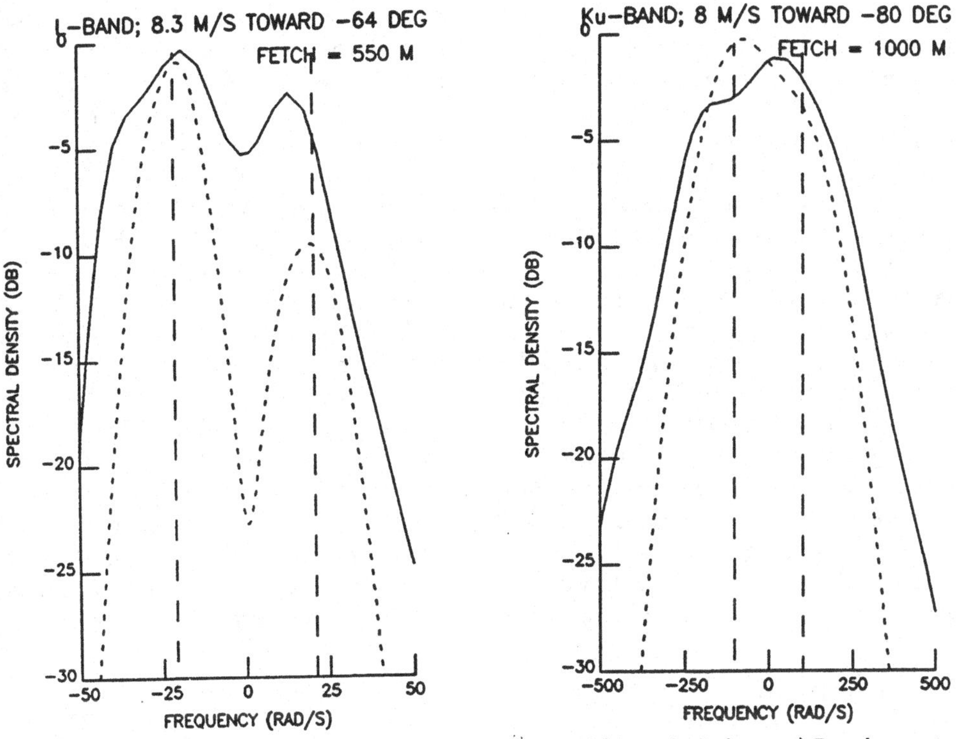

Figure 4: Comparison of measured (solid curves) an computed (short-dashed curves) Doppler spectra at L- and K_u-band. The vertical lines in the plots denote the surface-wave Bragg frequencies.

aircraft or satellite radars where the influence of surface-wave motion on the Doppler spectra is obscured by platform-motion effects. Very few low incidence angle measurements have been made from cliff- or tower-mounted radars due mainly to geometrical constraints.

More Doppler spectra measurements are available for incidence angles ≥ 45 deg [13,14], and these measurements are in general agreement with the $\theta_I = 50$ deg computations presented in Figures 2-3. In particular, the peak in these spectra are centered at the positive (negative) Bragg frequency for the upwind (downwind) radar look directions, and begin to show the presence of both positive and negative Bragg peaks as the look direction becomes more perpendicular to the wind direction. The width of the measured spectra increases with increasing wind speed; again in agreement with the computations.

The solid curves in Figure 4 show measured Doppler spectra adapted from [14] at L- and K_u-band; both for $\theta_I = 45$ deg. The wind velocity for the L-band measurement was 8.3 m/s toward -64 deg with respect to the radar look direction and the fetch was ≈ 550 m. For K_u-band, the wind velocity was 8 m/s toward -80 deg with respect to the radar and the fetch was 1000 m. We show by the short-dashed curves in each plot in Figure 4 our computed Doppler spectra for the corresponding conditions. We have accounted approximately for the limited fetch in both cases by lowering the wind speed used to determine $\psi(\vec{k})$ to that value which produces a dominant surface wave which has just enough time to develop over the given fetch. These wind speeds turnout to be 2.5 m/s for L-band and 3 m/s for K_u-band. (Note that we have not attempted to improve the agreement with the measured spectra by an adjustment of the wind speed.) It can see from Figure 4 that although the computed spectra are not as broad as the corresponding measured ones, they do reproduce the general features of the measurements. Namely, the L-band spectra show both the positive and negative Bragg peaks (denoted by the dashed vertical lines in the figure) with the negative peak being higher. The computed K_u-band spectrum is significantly broader than the L-band spectrum in agreement with the measurements. This broadening due to the long-wave motion obscures the Bragg peaks even though the wind is almost perpendicular to the radar. The degree of agreement between the measured Doppler spectra and our computations shown in Figure 4 is representative of that found for other cases in [14]. The most noticeable discrepancy is that the computations tend to underpredict the measured spectral widths. We feel that this discrepancy is probably not due to a deficiency in our scattering model, but rather due to our imprecise knowledge of the surface wave spectrum and how it varies as a function of wind speed and direction. Only with a better description of the surface - as well as accurate radar measurements - can we devise more stringent tests of the models.

4 Conclusion

We have presented in this paper calculations of Doppler spectra for microwaves scattered from the ocean surface. These calculations were made using a time-dependent composite model which is derived from an expression for the backscattered field correct through terms linear in the surface slope. Time dependence is included in this model by assuming linear evolution of each component of the surface-wave spectrum. Our model yields the proper small-perturbation fields for V-V and H-H polarization in the small-wave-height limit, and the proper (polarization-independent) specular fields when no small-scale roughness is present. We have shown that for low to moderate incidence angles, the Doppler spectra computed using our model are not only broadened considerably by the motion of the long-wave surface, but their peak frequencies are also shifted away from the Bragg frequency. The magnitude and direction of this shift (as well as the spectral width) are functions of the radar frequency, incidence angle, wind speed and direction.

At present, we cannot determine with certainty whether or not the assumptions inherent in the development of our scattering model are adequate. This is primarily due to the lack of detailed measurements of the 2-D surface spectrum concurrent with the radar backscatter data. In fact, there are not many low-incidence-angle Doppler measurements which we can compare with model predictions. At incidence angles $\geq 45\,\text{deg}$, some measurements do exist, and the general agreement with the predictions in this angular region is good. Accurate measurements of Doppler spectra are nevertheless needed over the full range of incidence angles with concurrent measurements of the wind speed and direction. When these data are collected - preferably along with direct measurements of the 2-D spectrum - we will be able to assess our models and determine the validity of the assumptions involved in their derivation. The completion of such an assessment will bring the possibility of obtaining quantitative remote measurements of the ocean surface much closer to reality.

ACKNOWLEDGEMENTS. It is a pleasure to acknowledge my colleagues J. R. Apel, R. F. Gasparovic, B. L. Gotwols, and J. R. Jensen for many useful discussions concerning this work. Also, I wish to thank W. C. Keller and W. J. Plant for permitting me to use their data prior to publication. This work was partially supported by the U. S. Office of Naval Research.

References

[1] Thompson, D. R., "Calculation of radar backscatter modulations from internal waves", to be published in JGR-Oceans, October, 1988.

[2] Holliday, Dennis, "Resolution of a controversy surrounding the Kirchhoff approach and the small perturbation method in rough surface scattering theory", *IEEE Trans. on Antennas and Propag.*, 35, 120-122, 1987.

[3] Stratton, J. A., *Electromagnetic Theory*, 615 pp., McGraw Hill Book Co., New York, 1941.

[4] Brown, G. S., "A comparison of approximate theories for scattering from rough surfaces", *Wave Motion*, 7, 195-205, 1985.

[5] Holliday, Dennis, Gaetan St-Cyr, and Nancy E. Woods, "Comparison of a new radar ocean imaging model with SARSEX internal wave image data", *Int. J. Remote Sensing*, 8, 1423-1430 1987.

[6] Bass, F. G., and I. M. Fuchs, *Wave Scattering from Statistically Rough Surfaces*, 525 pp., Pergamon Press, New York, 1979.

[7] Barrick, D. E., "Rough surface scattering based on specular point theory", *IEEE Trans. on Antennas and Propag.*, AP-16, 449-454, 1968.

[8] Valenzuela, G. R., "Theories for the interaction of electromagnetic and oceanic waves - A review", *Boundary Layer Meteorol.*, 13, 61-85, 1978.

[9] Thompson, D. R., B. L. Gotwols, and R. E. Sterner, "A comparison of measured surface wave spectral modulations with predictions from a wave-current interaction model", to be published in JGR-Oceans, October, 1988.

[10] Bjerkaas, A. W. and F. W. Riedel, "Proposed model for the elevation spectrum of a wind-roughened sea surface", Report TG-1328, The Johns Hopkins University Applied Physics Laboratory,1979 [NTIS ADA083426].

[11] Donelan, Mark A., and Willard J. Pierson, Jr., "Radar scattering and equilibrium ranges in wind-generated waves with application to scatterometry", *J. Geophys. Res.* 92, (C5), 4971-5029, 1987.

[12] Masuko,Harunobu, Ken'ichi Okamoto, Masanobu Shimada, and Shuntaro Niwa, "Measurement of microwave backscattering signatures of the ocean surface using X-band and K_a-band airborne scatterometers", *J. Geophys. Res.* 91 (C11), 13065-13083, 1986.

[13] Wright, J. W., "Detection of ocean waves by microwave radar; the modulation of short gravity-capillary waves", *Boundary Layer Meteorol.*, 13, 87-105, 1978.

[14] Plant, W. J. and W. C. Keller, "Evidence of Bragg scattering in microwave Doppler spectra of sea return", submitted for publication to JGR-Oceans, 1988.

LIMITATIONS OF THE TWO-SCALE THEORY FOR MICROWAVE BACKSCATTER FROM THE OCEAN

P. L. C. Jeynes
Oxford Computer Services Ltd
52 St Giles'
Oxford
OX1 3LU U.K.

Radar imagery of the sea surface reveals both surface features (e.g. wind structure, swell waves) and sub-surface features (e.g. internal waves, tidal flows over submerged sand banks). In addition appear a variety of man made objects (e.g. rigs, vessels) which may be sensed directly or via their effects on the oceanic environment (i.e. their wakes). An important part in the proper interpretation of such images is a thorough understanding of the scattering of microwaves from the sea surface. The other essential part of the problem is the hydrodynamics of the short wind waves and their modulation by the currents induced by surface and internal features. In the absence of wind the ocean is glassy-smooth giving trivial specular mirror scattering, largely irrespective of the presence of oceanic features. It is thus important to understand first the scatter from a wind-generated rough ocean surface (wind scatterometry) and later, the scatter in the additional presence of ocean features.

The three contending rough surface scattering theories are:

1. Crombie (Bragg) 1st order mechanism
2. Incoherent 2-scale composite surface model
3. Isakovich (Kirchoff) integral equation approach.

Detailed formulae and nomenclature are appended.

The Crombie (Bragg) mechanism has been extensively used for interpreting ocean radar imagery (Vesecky and Stewart, 1982). However it is a first order theory limited to slightly rough surfaces on the wavelength scale and is therefore at microwavelengths valid in only exceptionally calm conditions (Holliday 1986, Jeynes 1986). For Seasat SAR (L-band) the ocean roughness needs to be less than 2.5 cm, i.e. the windspeed less than a few metres per second, (these conditions are quite rare).

The 2-scale model was developed to account for the presence of large scale surface roughness supposing the sea to consist of slightly rough patches tilted by the larger, longer sea waves. In this the ocean is split into large and small scale roughness which requires the surface spectrum to be divided about some wavenumber into long and short wave components. There is a trade-off between the validity of the small perturbation Crombie scatter mechanism which requires the "tilted patches" to be only slightly rough and hence of small dimen-

G. J. Komen and W. A. Oost (eds.), Radar Scattering from Modulated Wind Waves, 41–47.

sion, against the requirement that the large scale surface be undulating so that diffraction from finite sized patches, and interference between neighbouring patches is negligible in comparison with the geometric effect of the large scale tilts. Many workers have claimed considerable success for this model, (e.g. Plant, 1986). However recent work (Jeynes, 1988) suggests the 2-scale model gives only an approximate description for calm to light sea conditions. Theoretical arguments based on a Phillips K^{-4} or a Toba $K^{-3\frac{1}{2}}$ spectrum both suggest an optimum scale division wavelength around ten radar wavelengths, $L_1 = 10\lambda$. The common place fine tuning of the spectral division parameter, or the parameters in the wind-sea spectrum to fit data, is probably unjustifiable. Further, the 2-scale models get worse in higher winds, though precise details depend on the form of the wind-sea spectrum, which is not well known at present.

Recently published 2-scale calculations by Donelan and Pierson (1987) show incorrect trends in the model output, such as

(i) $\sigma^{VV}(\lambda)$, slowly increasing from X to L-band with λ rather than decreasing according to data and intuition, see Figure 1, and

(ii) flat L-band $\sigma^{VV}(W)$ curves for $W = 2\text{-}40$ m/s, see Figure 2.

Donelan and Pierson choose their scale division wavelength as $L_1 = 40\lambda$ by fitting the 2-scale model to circle flight data set at K_u-band. However, because the 2-scale model is only very approximately valid for windseas then one cannot find L_1 by data fitting. The choice $L_1 = 40\lambda$ gives a small scale roughness height which does not satisfy the slight roughness criterion and so the model neglects dominant higher order perturbation terms which are larger than the 1st order Crombie (Bragg) term.

The use of the integral equation approach, in particular the 1st iteration (i.e. tangent-plane-approximation/Kirchoff theory) developed by Isakovich (1952) is suggested as the most promising way forward for wind scatterometry and imaging ocean features. Some of this important program of work has been started in the U.S. by Holliday and more recently Thompson. The comprehensive computation of ocean scatter as a function of the windfield and radar parameters using the Isakovich formula is regarded as a priority in the present program of work. Calculations using a variety of recently proposed ocean spectra should be performed. Comparison with 2-scale model calculations and sea truth backscatter data is of course essential and will surely suggest future directions of work. Backscatter data is widely dispersed in the literature and its amalgamation into a reference text or data tape would be of considerable value.

The 1st iteration is known to give no polarization effects which are commonly observed in backscatter from the sea. Further theoretical work on the 2nd iteration or possibly a "curved surface approximation" (Jeynes, 1986) is required before one can calculate polarization effects using present computing capabilities. The role of multiple scattering in depolarization from the ocean needs further investigation, in particular in anticipation of planned HV backscatter and image data.

Thus far we have considered only the average backscatter cross-section, for which the above theories have been developed. Now, the statistical distribution of ocean backscatter (related to SAR image speckle) is known to be non-Rayleigh and to fit a K-distribution rather better. Theoretical work and possibly Monte-Carlo simulations may lead to an understanding of the information the speckle statistics are conveying about the nature of the surface.

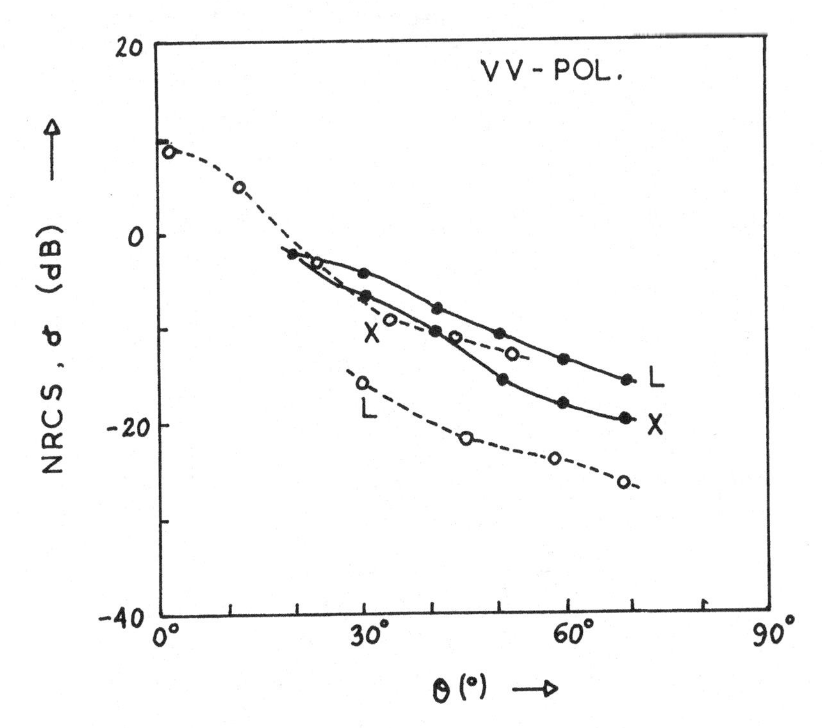

FIGURE 1. Problems in the comparison of 2-scale theories with sea-scatter data. The data are of vertical polarisation at X and L bands for roughly equal moderate to high windspeeds. The recent 2-scale model of Donelan and Pierson (1987) is reasonable at X-band but breaks down at L-band. An even worse problem: the underlying trend of the model on increasing the radar wavelength (at a given windspeed) is to increase the backscatter. This is contrary to intuition as a longer radiation wavelength will see a smoother surface and so tend to scatter less diffusely, thus we should expect less backscatter (away from vertical) as is shown in the data.

	band	λ(cm)	W(m/s)	
DATA --○--	X	2.2	15	Jones, Shroeder and Mitchell (1977)
	L	25	19	Daley (1973)
THEORY —●—	X	3	15	Donelan and Pierson's (1987) 2-scale model
	L	23.5	19	(extracted from their figs.32&34)

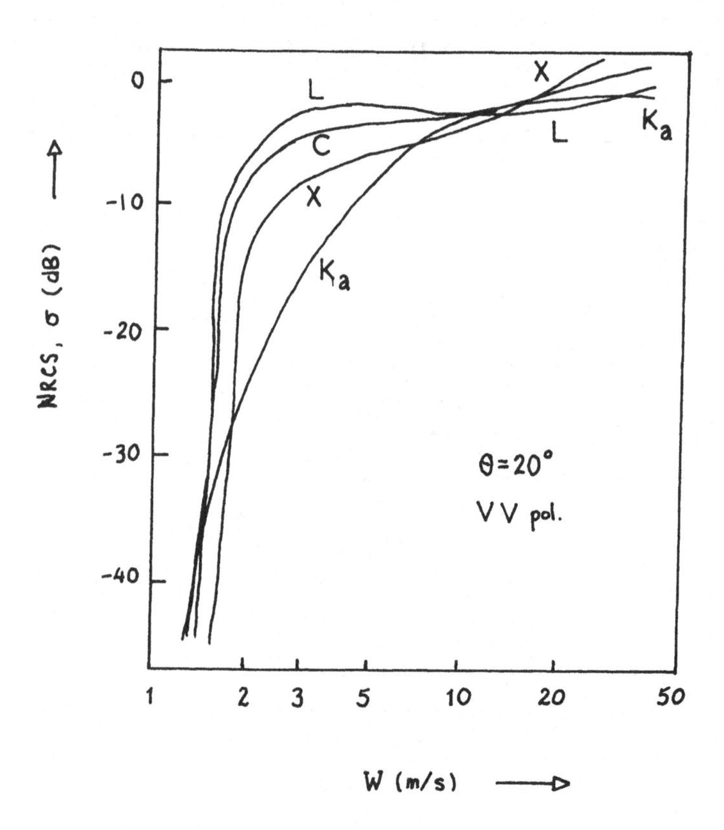

FIGURE 2. Further problems with Donelan and Pierson's (1987) results. The 2-scale model NRCS predictions for vertical polarization and θ=20° have been extracted from Donelan and Pierson (1987, Figures 32-35) for sea temperature 0°C. The light to moderate windspeed (2-12 m/s) behaviour is quite extraordinary. The consistent increase in backscatter with increasing wavelength is counter-intuitive (longer waves should "see" a smoother surface) and unsupported by data. The L-band backscatter appears almost constant over the range W=2 to 30 m/s contrary to much SAR imagery of wind structure.

band	L	C	X	K_a
λ (cm)	23.5	5.66	3	0.87

APPENDIX: ROUGH SURFACE SCATTERING THEORIES

1. Crombie (Bragg) mechanism

Backscatter NRCS (to 1st order in η)

$$\sigma_{\substack{VV\\HH}}^{(1)} = 16\ \pi\ k^4(1 \pm \sin^2\theta)^2\ \Psi(-2k\sin\theta, 0)$$

validity $h \lesssim \lambda/4\ \pi\cos\theta$ (i.e. slightly rough)

2. Two-scale incoherent model

Large and small scale surface ($\eta = \eta_\ell + \eta_s$) spectral scale division wavenumber $K_1 = 2\pi/L_1$

$$\sigma = \iint d\eta_{\ell x} d\eta_{\ell y}\ p(\eta_{\ell x}, \eta_{\ell y}) \sqrt{1 + \eta_{\ell x}^2 + \eta_{\ell y}^2}\ \left(\underset{\substack{\uparrow\\ \text{specular}}}{\sigma^{(0)}} + \underset{\substack{\uparrow\\ \text{Crombie}}}{\sigma^{(1)}}\right)$$

validity
- $h_s \lesssim \lambda/4\pi\ \cos\theta$ small scale roughness
- large scales $\gtrsim 10\lambda$ for approx. validity
- (Same as the criterion for the large scale surface alone to be a geometric optics scatterer.)

3. Integral equation approach

1st Isakovich (Kirchoff) iteration: tangent plane approximation. (Closely related to Huygens' principle.)

$$\sigma = \frac{q^4}{4\pi q_z^2} \iint dxdy\ e^{-i\underline{Q}.\underline{x}}\ e^{q_z^2 (\psi(\underline{x}) - \psi(o))}$$

validity
- extremely difficult to quantify - but we need not go all the way to the geometric optics limit, since includes diffraction, but not perfectly.

remarks
- no polarization effects
- reduces (in limit $h < \lambda/4\pi\cos\theta$) to Crombie scatter without the $\pm\sin^2\theta$ polarization term

2nd iteration: gives polarization effects and correct 1st order Crombie scatter.

NOMENCLATURE

L	ocean wavelength
K	ocean wavenumber ($2\pi/L$)
λ	radar wavelength
k	radar wavenumber ($2\pi/\lambda$)
L_1	2-scale model division ocean wavelength
W	windspeed
σ	normalized radar cross-section (NRCS)
$\underline{q}$	change in radar wavevector on scattering
$\underline{Q}$	horizontal projection of $\underline{q}$
V,H	vertical, horizontal polarizations
h	r.m.s. height
η	ocean surface height
Ψ	ocean wave spectrum
θ	angle of incidence
ψ	surface correlation function
η_ℓ, η_s	large, small scale surface
$\eta_{\ell x}, \eta_{\ell y}$	large scale slopes
$p(,)$	probability distribution of large scale slopes

ACKNOWLEDGEMENTS

It is a pleasure to thank Dr J.O. Thomas for his encouragement in the present work. This work was funded by the Procurement Executive, MOD. We thank Dr J.C. Scott, ARE Portland, for his support.

REFERENCES

Daley, J.C., 1973. 'Wind dependence of radar sea return' J. Geophys. Res. 78, (33) 7823-7833

Donelan, M.A. and Pierson, W.J., 1987. 'Radar scattering and equilibrium ranges in wind-generated waves with applications to scatterometry' J. Geophys. Res. 92 (C5) 4971-5030.

Holliday et al. 1986. 'A radar ocean imaging model for small to moderate incidence angles' Int. J. Remote Sensing 7 (12) 1809-1834.

Isakovich, M.A., 1952. 'Wave dispersion from a randomly uneven surface' Zh. Eksp. i. Teor. Fiz. 23, 305-314.

Jeynes, P.L.C., 1986. 'The scattering of electromagnetic radiation from irregular surfaces' Inst. Math. and Applus. Conf. Proc., Chelmsford May 1986 Oxford Univ. Press.

Jeynes, P.L.C., 1988. 'The microwave backscatter from the ocean - polarization effects, Part III: Theories cont'd' OCS Ltd. Report January 1988.

Jones, W.L., Schroeder, L.C. and Mitchell, J.L., 1977. 'Aircraft measurements of the microwave scattering signature of the ocean' IEEE Trans. AP 25(1) 52-61.

Plant, W.J., 1986. 'A two scale model of short wind-generated waves and scatterometry' J. Geophys. Res. 91(C9) pp10735-10749.

Vesecky J.F. and Stewart, R.H., 1982. 'SAR observations of the ocean surface' J. Geophys. Res. C-87, 3397-3430.

FIRST RESULTS OF THE VIERS-1 EXPERIMENT

D. VAN HALSEMA[1], BERND JÄHNE[2],
W. A. OOST[3], C. CALKOEN[4], P. SNOEIJ[5]

ABSTRACT. In February 1988, combined measurements of microwave backscatter, wind, waves and gas exchange have been carried out in the large Delft Hydraulics wind/wave tank. This experiment was the first in a series of experiments in the frame of the VIERS-1 project. In this project a number of Dutch and German laboratories cooperate. Main objective is to come to a physical description of the processes involved in wind scatterometry and, from that point, to an improvement of the algorithms used for determination of wind speed and direction from satelliteborne microwave scatterometers. A second objective is to study the relation between the gas exchange at the water surface and the microwave backscatter. To achieve these objectives two wind/wave tank experiments and one ocean based platform experiment are scheduled. In this paper, the VIERS-1 programme will be outlined. The Delft wind/wave tank experiment will be described and some first results of a preliminary comparison of backscatter and wave slope measurements will be shown.

1 The VIERS-1 Programme

After a long history of ocean tower based experiments [*De Loor*, 1983], we felt that the next step forward in understanding the microwave backscattering could only be a multidisciplinary study under controllable circumstances. So, in 1986, the VIERS-1 (Dutch acronym for "Preparation and Interpretation of ERS-1 data") started with the preparation of two wind/wave tank experiments and an ocean tower experiment. In the VIERS-1 team all disciplines involved in wind scatterometry are present: meteorology, oceanography, microwave technology, and microwave remote sensing.

Experiments in wind/wave tanks have one important advantage over ocean environment experiments: the conditions of wind and waves can be controlled and measured in detail. This especially counts for the measurements of small scale waves. This was the main reason for returning to wind/wave tanks for microwave backscattering research. Still one has to remember that a wind/wave tank is an artificial surrounding for studying microwave backscattering and that some scaling effects may occur. With the VIERS-1 programme, we are trying to eliminate these effects by using a stepwise approach:

[1]Physics and Electronics Lab TNO, P.O.Box 96864, 2509 JG The Hague, The Netherlands
[2]Scripps Institution of Oceanography, La Jolla, CA 92093-0212, USA
(on leave from Heidelberg University, West Germany)
[3]KNMI, P.O.Box 201, 3730 AE de Bilt, The Netherlands
[4]Delft Hydraulics, P.O.Box 152, 8300 AD Emmeloord, The Netherlands
[5]Delft University of Technology, P.O.Box 5031, 2600 GA Delft, The Netherlands

G. J. Komen and W. A. Oost (eds.), Radar Scattering from Modulated Wind Waves, 49–57.

Dimensions tank L × W × D [m]	100 × 8 × 0.8
Wind velocity range	0–15 m/s
Wave generation	regular and irregular waves, adjustable spectrum
Maximum wave height	0.3 m
Water circulation	< 1 m^3/s

Table 1: Some features of the Delft wind/wave tank

- First, an experiment in a large indoor, well equipped wind/wave facility of Delft Hydraulics in Delft;
- Second, an experiment in the much larger Delta tank where full-scale oceanic waves can be generated;
- Third, an ocean-based platform experiment from the Noordwijk tower, located 10 km off the Dutch North sea coast.

The Delft facility is excellently suited for a detail study of the processes of microwave backscattering. Wind and waves can be controlled and measured in great detail. However, the mechanically generated waves have a limited length and amplitude and wave age, because of the shallow water depth of only 0.8 m. Therefore, this tank was mainly used for studying wind-generated waves. The Delta tank with 5 m water depth and 260 m length allows to generate almost real-sized oceanic waves, much better suited for studying the effects of long waves on short gravity and capillary waves and microwave backscattering. This experiment is scheduled for March 1989. Finally, the Noordwijk tower experiment will be used as a final check on the validity of the tank experiments with respect to the possible influences of scaling effects.

Along with the increase in size of the facility, the measurement of small scale waves will become increasingly difficult. For each of the facilities, new imaging measurement techniques were and will be developed to work as close to the scatterometer footprint as possible.

2 The Delft Wind/Wave Tank Experiment

In February/March 1988, the first VIERS-1 wind/wave tank experiment took place in the Delft Hydraulics wind/wave tank. Refer to Table 1 for some of the characteristic features of this tank. The main objective of this experiment was to get a better understanding of the processes involved in microwave backscattering of the water surface. Wind, water surface, and microwave backscatter were measured under various circumstances. In this section attention will be paid to the diverse measurements and measurement techniques.

2.1 Microwave Backscatter Measurements

Microwave backscatter measurements were performed with a FM/CW X-band scatterometer. Some of the main characteristics of this scatterometer are listed in Table 2. The scatterometer was especially adapted for the wind/wave tank experiment in order to get a reasonable footprint at short distance from the antenna. A footprint of approximately

Type	FM/CW scatterometer
Frequency	9.6 GHz (X-band)
Frequency modulation	triangular, 50 Hz, 300 MHz
Polarization	HH or VV
Range resolution	50 cm
Antenna	Parabolic, 1.1 m diameter
Footprint	1×1 m^2
Phase error over footprint	$< 15°$
Azimuth angle range	0–180°
Incidence angle range	24.5–60°

Table 2: Specifications of the X-band FM/CW scatterometer

80×80 cm was obtained at 4.22 m distance from the antenna plane. Special attention was paid to get a flat phase front and antenna pattern over the whole footprint. This was achieved by using a 1.1 m diameter parabolic antenna with a miniature feed, operated in the Fresnel region [*Snoeij et al.*, to be published].

Incidence angle and azimuth angle could be varied over a wide range: azimuth 0–180° with respect to the wind direction; incidence over 24.5–60.0°. For all combinations of incidence and azimuth angle, the footprint was centered at 100 m fetch.

The frequency of the emitted signal from the scatterometer was triangularly modulated, with an amplitude of 300 MHz$_{pp}$ and a repeat cycle of 20 ms. A range resolution of 50 cm could be obtained. The high range resolution proved to be necessary in order to avoid reflexes from the tank entering the signal.

2.2 Wave Measurements

A number of wave measurement devices have been employed in the Delft wind/wave tank during this experiment. The most important for the microwave measurements was the so-called imaging slope gauge (ISG). The principle of this technique has been described in more detail elsewhere [*Jähne* 1988; *Keller and Gotwols*, 1983]. The ISG consists of a submerged illumination source in the bottom of the wind/wave tank and a camera at the ceiling looking onto the water surface. The light source illuminates the water surface with a graded light intensity. As a consequence, the gray level recorded by the camera is directly related to the slope of the waves. By means of a calibration procedure, the absolute slope of the waves can be determined over the whole imaged area, with a repeat cycle of 25 Hz. The illumination gradient can be switched in either along-wind or cross-wind direction. In this way, either the along-wind or the cross-wind slopes can be visualized. The dimension of the imaged area can be varied by means of a zoom lens up to a maximum area of about 1×1 m. The illumination source is located directly beneath the scatterometer footprint, so simultaneously measurements of backscatter and wave slope images can be taken at the same footprint. The synchronization of scatterometer and ISG data is within 20 ms.

Besides the ISG, two other imaging wave measuring methods have been operated: the reflective slope gauge (RSG) and the reflective stereo slope gauge (RSSG). Both instruments are based on light reflection from the water surface. Since no part has to be submerged, they can also be applied to the field. The RSG was used to determine the two-dimensional slope distribution in real time. The RSSG offers the advantage to combine slope and height measurements in one instrument. The area imaged by it was approximately 30 × 40 cm. It was located close to the scatterometer footprint at approximately 98 meter fetch. All stereo pairs were immediately digitized and stored on magnetic tape for further processing.

Other wave measuring instruments operated were a laser slope gauge (LSG), described by *Lange et al.*, [1982] and *Jähne* [1988] and a wire wave gauge [*Lobemeier*, 1981]. These instruments provide well calibrated point measurements of the two-dimensional wave slope and the wave height. They will also be used to intercalibrate the other sensors.

2.3 Wind Measurements

The wind in the tank has been measured using a number of different sensors. The most valuable turned out to be a pressure anemometer [*Oost*, 1983]. This device measures the time-dependent three-dimensional wind vector. It is especially suitable for tank measurements because of its small measurement volume (only a few centimeters). Friction velocity and drag coefficient can be obtained from these data.

The 3-D wind vector was also measured with a sonic anemometer. Both instruments were operated close to each other on a transportable platform. Extensive measurements of the homogeneity of the wind field in the tank have been made.

The wind profile has been measured by means of an array of miniature cup-anemometers, at five different heights above the water surface. The cup-anemometers were mounted on a transportable platform and have been used at different positions across the end section of the tank.

2.4 Gas Exchange Measurements

The rate of gas exchange was measured using the controlled flux technique which — for the first time — allows an investigation of the local and instantaneous transfer velocity across the aqueous boundary layer. The new method is based on an infrared remote sensing technique [*Jähne et al.*, 1988]. In a test period in November 1987 also gas exchange tracer experiments have been carried out using ^{3}He and SF_6. The ^{3}He experiments were done in cooperation with P. Schlosser and K.O. Münnich from Heidelberg University, whereas R. Wanninkhof from Lamont Doherty Geological Observatory, Columbia University did the SF_6 experiments.

3 Some First Results

Measurements of the microwave backscatter have been performed, together and simultaneously with wave slope measurements as a function of azimuth angle, incidence angle, polarization and wind velocity. In this paragraph some very first results of these measurements will be shown to illustrate the potential of the database acquired.

Some typical results for the time series of microwave reflectance can be seen in Fig. 1a and Fig. 2. Figure 1a shows the, for the moment still uncalibrated, microwave backscatter at HH polarization for upwind conditions at an incidence angle of 45 degrees and at a friction

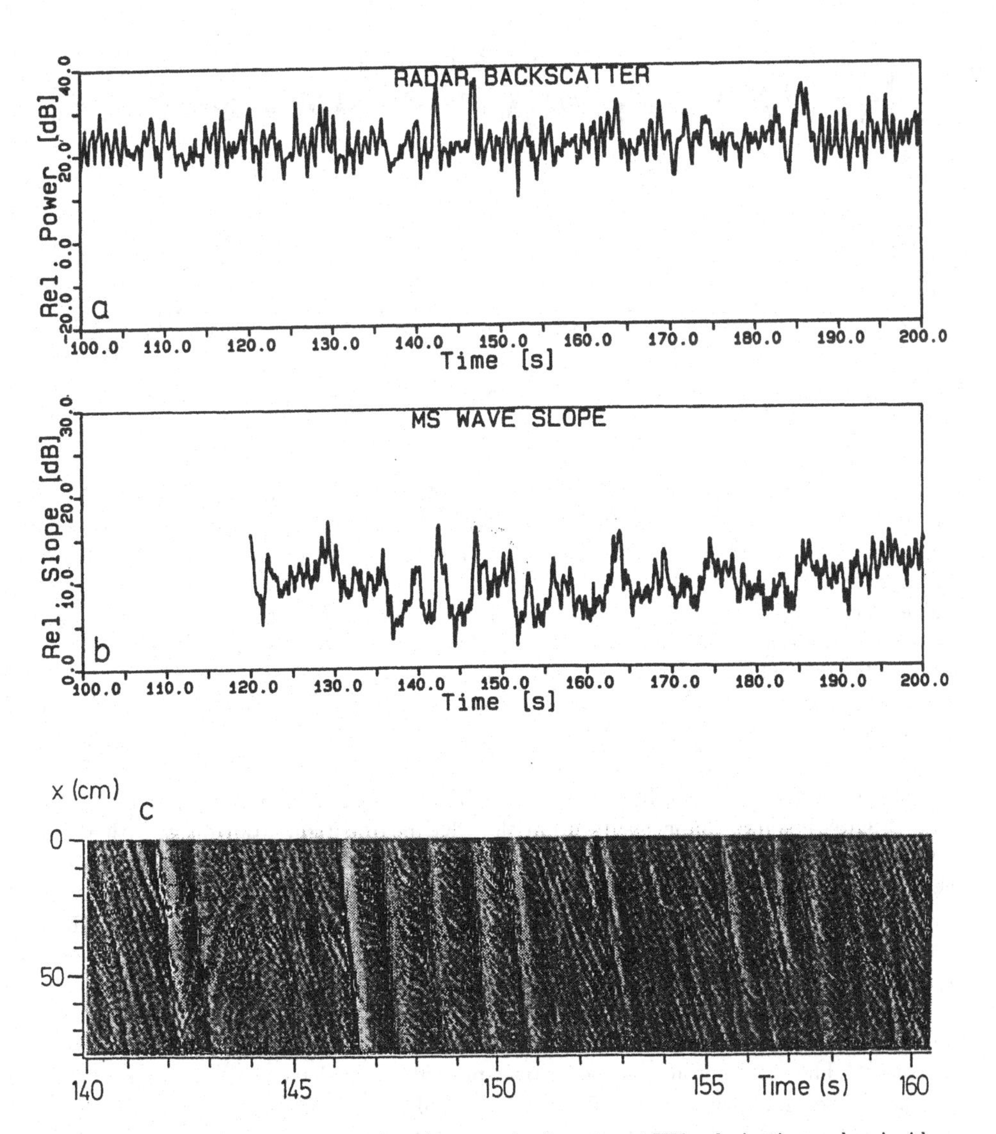

Figure 1: Simultaneous time series of the microwave backscatter at HH polarization and an incidence angle of 45°, upwind (a) and the mean square (ms) wave slope calculated from images recorded with the ISG (b). The friction velocity u_* was 0.37 m/s ($U_{10} = 11.5$ m/s). Note the strong modulation of the signals by the long waves (0.9 second period). The two spiky returns at $t = 142$ s and $t = 146.5$ s are caused by breaking waves. (c) shows a space-time image of the wave slope. One along-wind directed line in the center of the radar footprint was taken from the two-dimensional ISG images and a new image was formed showing the time dependency. The wave slope is coded as different intensities, where dark gray values indicate wave slopes in wind direction.

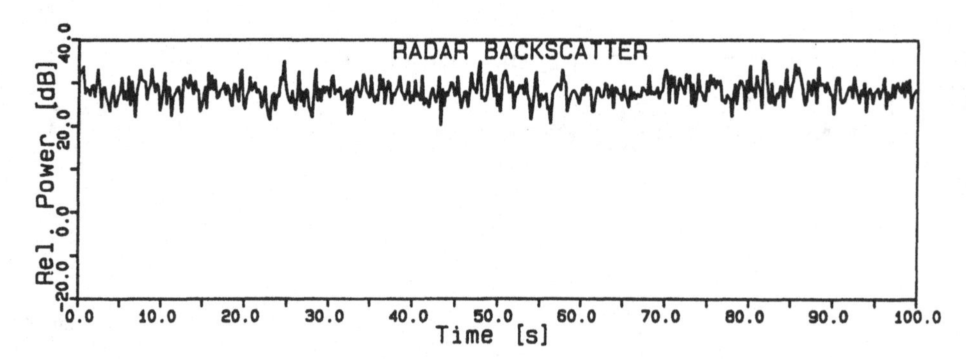

Figure 2: Same as Fig. 1a, only at VV polarization. The backscattered signal is much less modulated by the long waves.

velocity u_* of 0.37 m/s (U_{10} 11.5 m/s), and a fetch of 100 m. Fig. 2 was recorded under similar circumstances at VV polarization. The data has been averaged over 0.2 seconds (10 independent samples), which reduces the inherent Rayleigh speckle to ±2 dB (90% boundaries).

The longest wavelength at this wind speed is approximately 1 m and hence can just be resolved by the scatterometer. Modulations of the microwave backscatter by these waves in the HH polarized signals (Fig. 1) are clearly visible, whereas they are considerably lower at VV polarization. This effect is in qualitative accordance with platform measurements [*De Loor*, 1983]. At $t = 142$ s and $t = 146.5$ s in Fig. 1 two spiky returns occur of about +15 dB with respect to the background level.

By means of the ISG recordings these could be identified as breaking waves in the footprint. Figure 1c shows some results from the ISG, recorded simultaneously with the microwave signals. Although the ISG records time series of two-dimensional images, here only one particular line in the ISG image is displayed as a function of time for computational reasons. This line is from the center of the image, and lies in along-wind direction. Hence, it is in the propagation direction of the dominant waves. The gray values in the images of the ISG are related to the slopes of the waves. A dark gray value means a slope toward the propagation direction of the dominant waves, whereas a light one means a slope in the opposite direction. Zero slope at the crest or the trough of waves shows up as a intermediate gray values. The results from the wave slopes presented here have not been calibrated yet. Care should be taken when comparing signals recorded under different circumstances.

As waves travel through the footprint of the ISG, they show up as a dark band followed by a lighter band. A nice group of travelling waves can be observed in the middle of Fig. 1c. Waves travel from top to bottom in the space domain and from left to right in time. The finite travelling time through the footprint causes a deviation from a vertical line. The angle of this turning is a measure of the phase speed of the wave: the larger the angle, the smaller the phase velocity.

Also smaller waves can be observed. Because of their slower phase speed, they show a larger angle to the vertical than the dominant waves. See e.g. the region around $t = 155$ s. On top of strong dominant waves (e.g. $t = 150$ s) they are clearly modulated by these waves both in amplitude and phase speed. The modulation of the phase speed shows up in

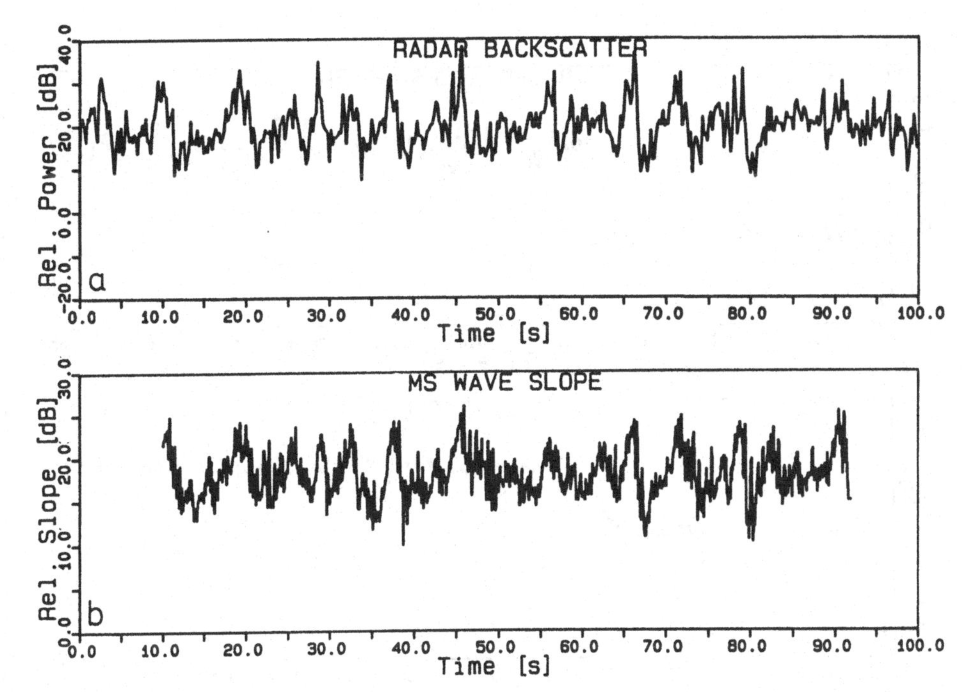

Figure 3: Time recording of the microwave backscatter at 35° incidence angle upwind and HH polarization and the ms wave slope as obtained from a single line from the imaging slope gauge. The friction velocity was 0.15 m/s (U_{10} = 5.8 m/s). Note the spiky behavior of the backscattered signal and the striking correspondence in long term variations of ms wave slope and microwave backscatter. The spikes are related to waves generated immediately after a micro-scale breaking event.

the S-shape bands across these longer waves.

At about t = 142 s and t = 146.5 s very steep slopes can be observed (white and black patches, close together). These belong to a breaking wave. This is the same breaking wave which showed up in the microwave backscatter signal (Fig. 1a). Note the smoother water surface in between the two breaking events.

By calculating the mean square slope over one line of the ISG image, one gets some measure of the surface roughness over that line and therefore over the footprint for that time. Care has to be taken, because one line is not always a good representative for the whole footprint, but as a first indication it is valuable. Figure 1b shows a time series of the logarithm of the mean square slope, calculated this way.

A lot of short term modulation with the frequency of the dominant wave can be seen (1.1 Hz). But also long term increases of the mean square (ms) slope occur with a duration of about 5 s. Although the ms wave slope is not the right wave parameter to be compared with the microwave backscatter measurements, it is very tempting to do so. If one does so, it is amazing to see how well the long term and also the short term variations of the microwave backscatter follow the variations of this mean squared slope.

Figure 3a shows a time recording of the microwave backscatter with the radar at an

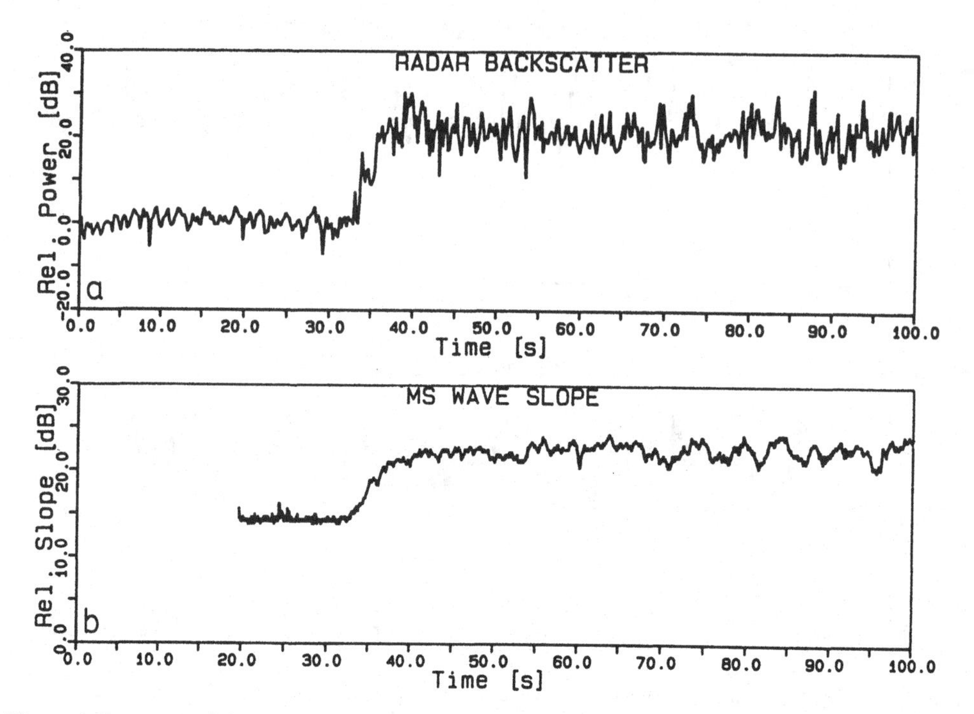

Figure 4: Response of the microwave backscatter (a) and the ms wave slope to the increase of wind speed of 0 to 7 m/s. Radar: HH polarization, incidence angle 45 degrees, upwind.

incidence angle of 35 degrees, upwind looking at HH polarization. The wind conditions were: $u_* = 0.15$ m/s, ($U_{10} = 5.8$ m/s), full fetch of 100 meter. At first glance, it is amazing how spiky the behavior of the backscattered signal is, especially considering the relative low wind speed, where one would not expect much breaking of waves. Increases and drops of the backscattered signal of over 15 dB occur regularly. Looking at the wave slope recordings, it appears that the regions of increased backscatter are caused by the strongly localized instabilities of the dominant waves (micro-scale wave breaking). Just after the breaking, circular waves are generated (like dropping a stone in a pool), which travel partly against the wind direction. The whole micro-scale breaking event increases the ms slope a great deal (Fig. 3b). The general trends for the radar backscatter and the ms wave slope fit extremely well.

Another interesting comparison between radar and ms wave slope is shown in Fig. 4. In this case, the wind speed was turned on to 7 m/s. Soon after the first waves appear, the radar responds immediately to the increased roughness on the water surface. The whole process of wave growth up to the equilibrium wave spectrum takes several minutes at this wind speed. Both the microwave backscatter and the ms slope reach an equilibrium value with a much faster time constant of approximately 10 s. A small overshoot occurs at about 6 s after the first waves were generated. This phenomenon cannot be traced back to the ms slope values. It has not been analyzed yet.

4 Summary and Conclusions

The VIERS-1 project was outlined. This project is aimed at a better physical understanding of the processes involved in microwave backscattering from the ocean. Main objective is to come to a physical description of these processes and, from that point, to an improvement of the algorithms used for the determination of wind speed and direction from satelliteborne scatterometers. A second goal is to study the relation between gas exchange and microwave backscattering. Two wind/wave tank experiments and one ocean based platform experiment are scheduled.

The first tank experiment took place in February/March 1988 and was described in this paper. Attention was paid to combined wind, wave, and microwave backscatter measurements together with gas exchange measurements. Microwave backscatter measurements have been made simultaneously and at the same footprint as imaging wave slope measurements. The data processing just started, but already the few examples presented here indicate that a very valuable data set has been acquired. The preliminary results of simultaneous microwave backscatter measurements and wave slope measurements at the same footprint clearly demonstrate the great potential for studying the mechanisms of microwave backscattering from the water surface.

Acknowledgements. The contribution of the Dutch laboratories was supported by the Netherlands Remote Sensing Board. The Institute for Environmental Physics, Heidelberg University was supported by a grant of the state Baden-Württemberg. The authors would like to thank Dr. P. Lobemeier for lending his wire wave gauge.

References

Jähne, B., Energy balance in small-scale waves: an experimental approach using optical slope measuring technique and image processing, this volume, 1988.

Jähne, B., P. Libner, R. Fischer, T. Billen, and E. J. Plate, Investigating the transfer processes across the free aqueous viscous boundary layer by the controlled flux method, *Tellus*, in press, 1988

Keller, W. C. and B. L. Gotwols, Two dimensional measurement of wave slope, *Appl. Optics, 22*, 3476–3478, 1983.

Lange, P. A., B. Jähne, J. Tschiersch, J. Ilmberger, Comparison between an amplitude-measuring wire and a slope-measuring laser water wave gauge, *Rev. Sci. Instrum., 53*, 651–655, 1982.

Lobemeier, P., A wire probe for measuring high frequency sea waves, *J. Phys. E. Sci. Instrum., 14*, 1407–1410, 1981.

Loor, G. P. de, Tower mounted radar backscatter measurements in the North sea, *J. Geophys. Res., 88*, 9785–9791, 1983.

Oost, W. A., The pressure anemometer - an instrument for adverse circumstances, *J. of Climate and Appl. Meteor., 22*, 2075–2084, 1983.

Session 3

The energy balance in short waves

Chairperson: W. A. Oost

EFFECTS OF REDUCED SURFACE TENSION ON SHORT WAVES AT LOW WIND SPEEDS IN A FRESH WATER LAKE

Kristina B. Katsaros[1], Hermann Gucinski* [2], Serhad S. Ataktürk[1], Robert Pincus[1]

1 Department of Atmospheric Sciences
University of Washington
Seattle, WA 98195

2 Anne Arundel Community College
Arnold, MD 21012

ABSTRACT. Measurements of the effects on wind waves by surface active compounds were made at the downwind end of a 7 km fetch in Lake Washington. Slicks produced from a model compound which reduced surface tension by 6-8 mN/m, formed 10 m patches which drifted by the wave sensor in 3-5 minutes. Naturally forming slicks, reducing surface tension by 1 to 2.8 mN/m, were also observed. For wind speeds of 2-3 m/s, waves of intrinsic frequency greater than about 2 Hz were strongly reduced, while at 6 m/s the reduction was weaker but still discernible. Depending on wind speed and slick type the reduction in spectral power density was 20 to 90% for gravity-capillary waves and up to 100% for capillary waves.

1. Introduction

The objective of our pilot study was to measure the effects on wind waves of slick forming substances similar to those found in nature, as a function of wind speed and wave frequency (or wave number), and to compare the results to the effects of naturally forming slicks.

Many of the active and passive remote sensing techniques employing microwaves depend on the presence of short ocean waves to interact with electromagnetic radiation (e.g., Ulaby, et al., 1982). Small waves produce constructive interference through Bragg scattering of radar signals. Emissivity at microwave wavelengths is a function of surface roughness, determined by waves similar in scale to the wavelengths of the electromagnetic radiation. As a result, interest in the effects of natural and man-made slicks on the short waves has increased.

It is everyone's experience that slicks modify the water surface by reducing the amplitude of short waves and creating a smooth looking patch. Scott (1979) lists many references on the calming effects of oil on water. The effects on radar backscatter and

---

* Present address: NSI/CERL, U.S. Environmental Protection Agency, Corvallis, OR 97333.

G. J. Komen and W. A. Oost (eds.), Radar Scattering from Modulated Wind Waves, 61–74.

emission include observations from aircraft (Hühnerfuss, et al., 1983a; Hühnerfuss and Alpers, 1983).

Laboratory investigations including those in water-wind tunnels have provided descriptive and quantitative data on capillary wave damping (Davies and Vose, 1965; Garrett, 1967; Hino, et al.,1969; Lin and Liu, 1979), and delineated the effects of surfactant films on the suppression of capillary wave onset (Gottifredi and Jameson, 1968; Scott, 1972). Explanations that relied only on differences in surface tension as the dominant parameter were inadequate and lead to a reassessment of the causes of wave damping (Lucassen, 1981). Theories have now been developed that incorporate both effects of surface tension and effects of surface dilational elasticity (the Marangoni effect) which is said to remove and redistribute energy between the small scale gravity and capillary waves (Hühnerfuss, et al., 1985 a,b; Hühnerfuss, et al., 1987). Scott (1972) reports that over some concentration ranges, surface to bulk diffusion can be expected to reduce dilational elasticity to produce the effects often observed.

Scott also investigated the differences in the combined effects of surface tension and dilational elasticity of low and high molecular weight surfactants. With oleic acid or oleyl alcohol, often chosen as representative compounds of low molecular weight, surface tension is strongly concentration dependent until a highly compressed monolayer is produced. On the other hand some high molecular weight compounds, such as polyethylene oxide, produce a smaller surface tension change that varies little with concentration. Results show wave damping at low wind speeds by such compounds to be comparable to that by oleyl alcohol, but at much higher surface tension values, just below the values of clean water.

This may be significant when considering wave damping from natural, biogenic slick forming products in open and coastal waters. It is rare that surface tension values in open ocean areas fall below clean seawater value by more than a few mN/m. Both lipid-like, low molecular weight compounds and high molecular weight polysaccharides and proteins or protein breakdown products occur at sea and may be concentrated at the sea surface (Baier, et al., 1974). Natural slicks are typically a mixture of molecules, often appearing to be dominated by those of greater molecular weight (Baier, et al., 1974; Gucinski, 1983; Carlson, et al., 1980). Mass balance considerations suggest that we can only account for 50% to 60% of the total organic carbon present as known species (Herr and Williams, 1987). It appears likely that the fraction still unaccounted for comprises the high molecular weight species, where humic acids and other refractory compounds make chemical analysis and speciation extremely difficult.

The question of how surface slicks affect small waves in a random wave field under a turbulent wind merits further study. Our field facility for measuring wavelengths as short as about 1 cm, using a single resistance wire (Ataktürk and Katsaros, 1987) combined with surface wind speed measurement allowed us to address several questions. Among these are: What sizes of waves are affected by "typical" surface slicks? At what wind speeds are the slicks so disrupted that their wave damping becomes minimal? We do not attempt to ascertain the effects of variations in slick composition, and size, although we expect that slick size may affect the low frequency cut-off of observable damping.

We studied mainly artificial slicks, which consisted of a model compound that could be applied repeatedly and was of consistent composition. Effects of naturally forming slicks were observed as opportunity arose at the same field site and with the same instrumentation (Ataktürk, 1988).

2. Materials and Methods

2.1. SITE AND INSTRUMENT DESCRIPTION

The field station, is located at Sand Point in Lake Washington. A bottom-mounted mast is positioned in 4 m water depth about 15 m from the shore. A 125 μm diameter resistance wire wave gauge is mounted on a horizontal boom 1 m above water level and 2 m from the mast (Figure 1). The frequency response of this wave gauge has been described by Liu, et al., (1982) and by Ataktürk and Katsaros (1987). For waves of 15 Hz frequency the response is 85% of the true wave amplitude. We have chosen this frequency as the cut-off for analysis in the present study. Work by Hühnerfuss, et al., (1987) and work cited by Hühnerfuss, et al., (1983b); Lobemeier (1978), strongly suggest that the presence of oils and surfactants do not affect wire wave gauge performance. A single horizontal Gill propeller anemometer was mounted 1.1m above the lake surface close to the wire wave gauge in order to have small time lag between the measured wind speed and wave height, a condition desired for the anticipated low wind speeds. The fetch was a maximum of 7 km for northerly winds. The data for the model slicks were collected during two days in August 1987, when the wind speed varied from 2 to 6 m/s. The naturally forming slicks occurred under similar environmental conditions.

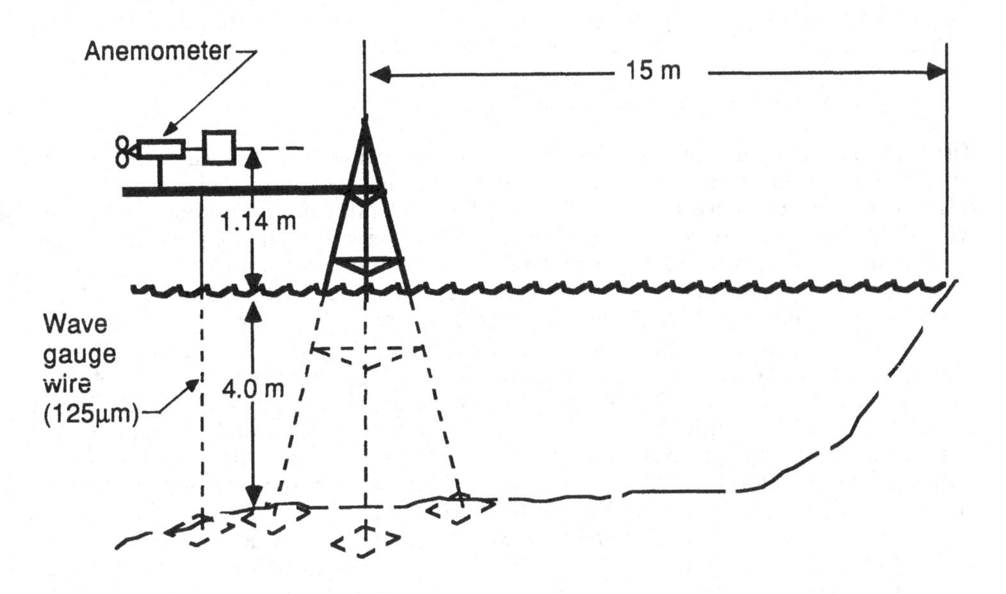

Figure 1. Sketch of MsMast (Mast for studies of Microscale Air-Sea Transfer) showing instrumentation.

2.2. SLICK PRODUCTION AND CHARACTERIZATION

Proteins may be used as model compounds for natural slicks because they possess high spreading pressures and appear as a significant constituent of natural films and slicks (Baier, 1974; Gucinski and Goupil, 1981), but their density promotes sinking and makes delivery a problem, while their solubility may influence slick composition when surface tension gradients become large. We therefore prepared a mixture of film forming surfactants containing protein (alpha-keratin, found in commercial gelatin), vegetable oil, having limited, but non-zero, spreading pressure and high buoyancy, and a commercial liquid detergent to act as a dispersant in order to enhance the initially slow spreading rates of the protein. Volume ratios were 3:3:2, respectively. We did not vary or attempt to optimize the composition.

Slow release of material from an initially capped jar, released by swimmer, produced slowly spreading, persistent, and reproducible slicks (size ~10 m) that drifted through the sensor area (duration 3~5 min). They had a sharp onset region at their leading edge and appear less well defined at the trailing edges.

The surface tension of the lake water was measured in the manner of Adam (1937), using non-spreading pure mineral oil modified by increasing concentrations of n-dodecyl alcohol, which gave a series of solutions of increasing spreading pressure. Drops of solution were applied to the surface in order of increasing spreading pressure until spreading against pre-existing surfactants was observed. This allowed estimates of surface tension in ranges chosen to be narrow for small deviations from clean, fresh water (72.4 mN/m at 20° C) and wider for large water surface tension changes.

Chemical signatures were obtained by Langmuir-Blodgett retrieval of surface material onto a Germanium prism subsequently analyzed by ATR (attenuated total reflection) infrared spectroscopy (Gucinski and Goupil, 1981).

2.3. ANALYSIS OF WAVE HEIGHT DATA

The wave height signal from the wire wave gauge was sampled in two ways. A low-pass filter selected frequencies < 50 Hz, while a high pass filter took the signal at frequencies > 6 Hz. The high pass signal was amplified by a factor of 110 before recording. In Figure 2 we see that overlap between the two channels from the wave gauges shows very good agreement. These two wave gauge signals, the wind-speed signal and a running commentary in a voice channel were recorded on a Hewlett Packard FM multichannel tape recorder. The analog tape data were digitized at a rate of 128 samples per second and written to computer compatible tape by a Raytheon 704 computer. The signals were displayed on high speed recorder charts. Records of measurements in and outside of slicks were selected by inspection of the high-pass wave data in conjunction with the voice commentary. A Fast Fourier Transform was applied to either 45 seconds or 90 seconds of record, depending on the duration of the slick at the wave gauge. After application of a Hanning filter, spectra were calculated using 1024 points, corresponding to 8 seconds of the time series. Eight such records were averaged and the results were smoothed over 3 adjacent frequencies to produce the final spectral densities. The frequency range for which acceptable measurements can be made ranges from about 0.2 to 15 Hz. For a slick covered surface, however, the amplitude of waves at frequencies greater than 10 Hz become less than the resolution of the wave gauge wire and the power spectrum flattens out.

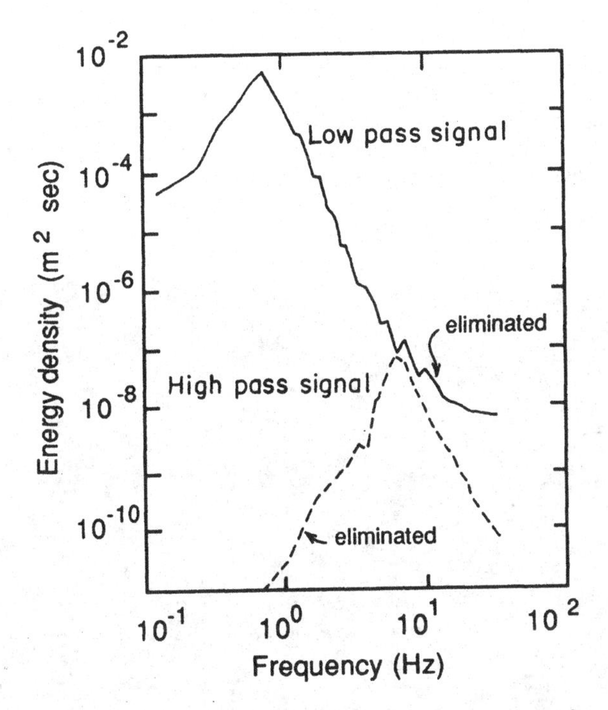

Figure 2. Sample spectra that show the effect of wave gauge performance.

The change in spectral densities from clean to contaminated water surface was calculated as:

$$\text{per cent change at a given frequency} = \frac{\Phi_{no} - \Phi_{slick}}{\Phi_{no}} \times 100\%$$

where Φ_{no} = pre- or post-slick spectral power density

Φ_{slick} = spectral power density in the slick in the slick

For the model slicks, correction for dynamic response of the wire or for Doppler shift in frequency of the small waves due to orbital velocity of long waves (Ataktürk and Katsaros, 1987) were not performed. At the lowest wind speeds (2.6 m/s) this is justified, since any Doppler shift is small and ill-defined. At the higher wind speeds (6.0 m/s) the frequencies reported may differ significantly from the intrinsic frequency of the wave. However, the gravity waves,which create Doppler shifts, did not differ measurably within and outside the slicks, hence the additional effort to correct the spectra was not deemed necessary. In case of the naturally formed slicks, which drifted by the station, the Doppler shift correction was part of the routine analysis. Once the intrinsic frequency spectra are obtained, wave number spectra can be calculated from the dispersion relation (e.g., Kraus, 1972). In this calculation the surface tension value for clean water was used for both clean and contaminated water. This would only lead to a small error in the wave number since surface tension of naturally formed slicks seldom differ from that of clean water by more than 10%.

3. Results

Artificial slicks produced for this study and the often observed "natural" slicks left prominent visual signatures. Figure 3 shows a slick at the wave gauge. Typical duration of slick passage, a function of its size as well as its velocity resulting from wind stress and surface currents, was from 3 to 5 minutes. This limits the sampling interval and therefore the lowest frequency measurable for spectral analysis.

Figure 3. Visual capillary wave damping by slick.

3.1. MODEL COMPOUND SLICKS

The conditions during measurements of the effects of model compound slicks are listed in Table 1. The surface tension in these slicks measured with the oil drop method was about 68 mN/m.

TABLE I: Environmental Conditions during Model Slick Runs

Date	Run	Duration (s)	T(dry) °C	T(wet) °C	T(water) °C	Wind m/s	Notes
17Aug	1	66	21.4	16.8	22.4	5.93	
pm	2	72				6.31	
	3	44				6.51	waves break
18 Aug	4	76	17.4	14.8	20.6	2.64	
am	5	84				2.59	
18 Aug	6	48	20.8	15.6	≈22	5.95	waves break
pm	7	52				5.38	
	8	88				5.39	

Figures 4 a and b show pre- and post-slick as well as slick affected spectral densities at 2.3 m/s and 5.9 m/s wind speeds. Note that in both cases the pre- and post-slick spectra are in close agreement, which illustrates the precision of each spectral estimate.

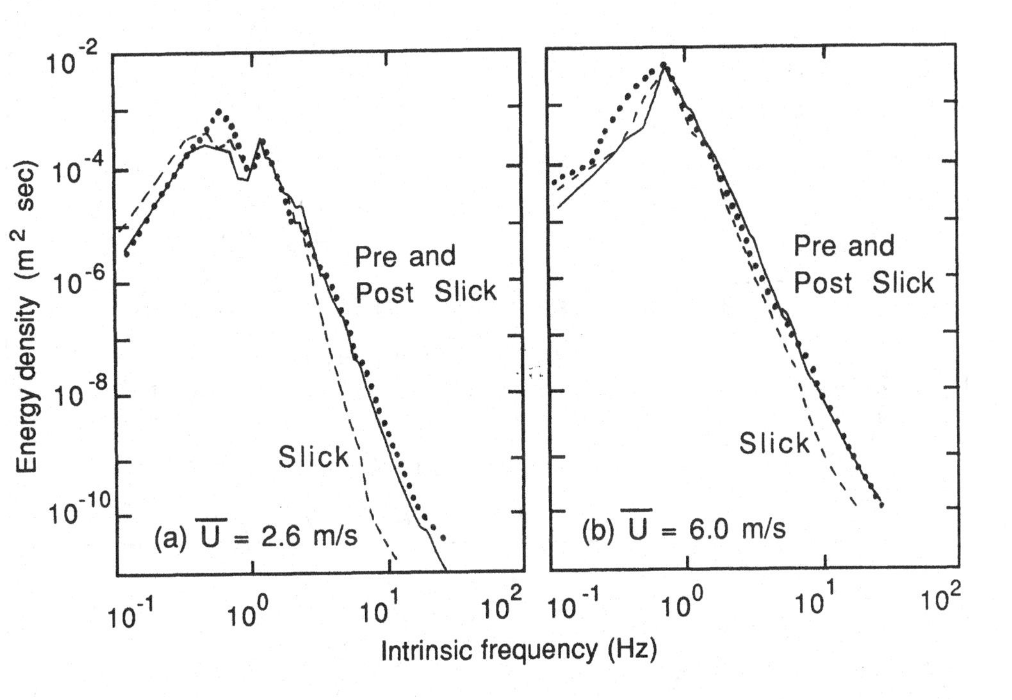

Figure 4. Wave amplitude spectra in "clean" and slick covered lake water for two mean wind speeds, $\bar{U}$.

The per cent change for two cases is shown in Figure 5. The curves were generated from the equation for the smoothed slopes of the spectra, chosen because they contained little noise. In the cross-hatched region beyond 10 Hz the results are in error; because of the low signal from the slick-covered surface only noise remains.

3.2. NATURALLY FORMED SLICKS

Analysis of an infra-red spectrum of dominant molecular bonds at the air-water interface, made from a germanium prism dip, indicates that the naturally-formed slicks consist of biogenic materials with petroleum residues also present (Figure 6).

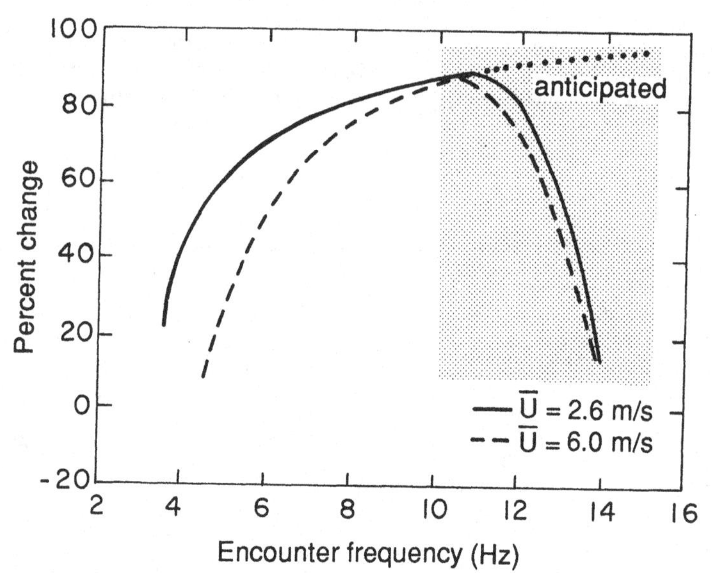

Figure 5. Per cent change in spectral power density as a function of encounter frequency for model slicks. In the cross-hatched region the amplitude is less than the resolution of the wave wire.

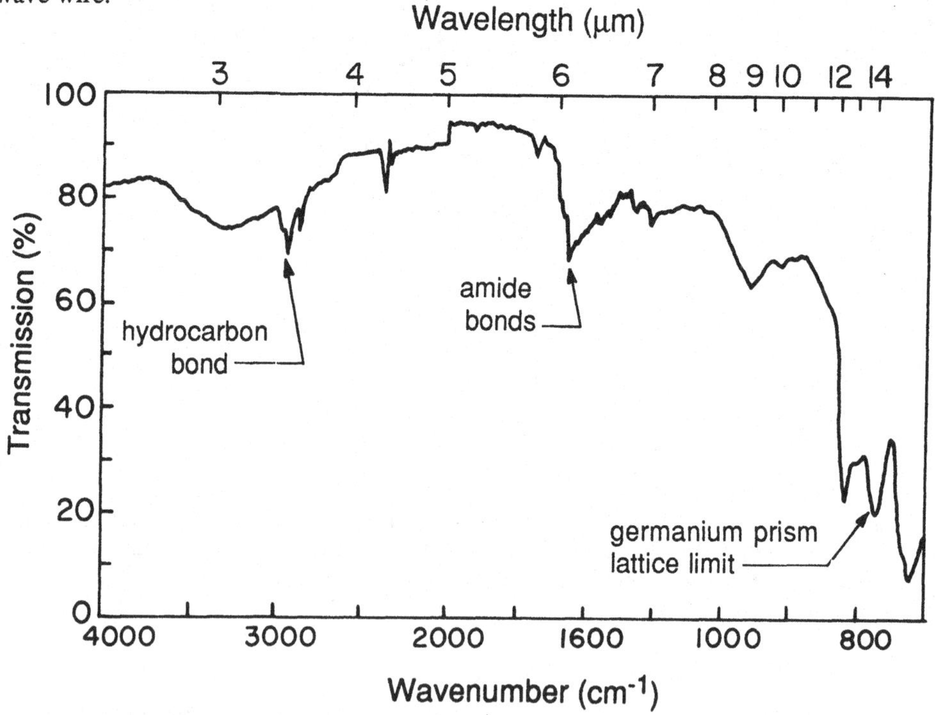

Figure 6. ATR-infrared spectrum of dominant molecular bonds at the air-water interface in a naturally formed slick on Lake Washington.

Considerable recreational boat use, and limited commercial shipping are likely sources for the man-made inputs. Adam spreading oils suggested that deviations of surface tension of the lake water from that of clean water were small, typically 1 to 2.8 mN/m. This implies that this type of slick can have considerable effects (as will be seen below) in reducing the spectral power density at low surface tension depressions. Table 2 lists the observed environmental conditions for these slicks, while Figure 7 shows the wave number (intrinsic frequency) spectra for two slicks at slightly different wind speeds. The higher wind speed case shows the stronger effect of the slick but since the two slicks may not have been identical in composition these results do not necessarily contradict the model slick findings.

TABLE II: Environmental Conditions during Runs with Naturally Formed Slick

Date	Run	Duration (s)	T(dry) °C	T(wet) °C	T(water) °C	Wind m/s
8/25/86 pm	NF 1	240	24.0	20.0	23.5	3.5
8/27/86 pm	NF 2	180	27.0	20.5	24.5	2.4

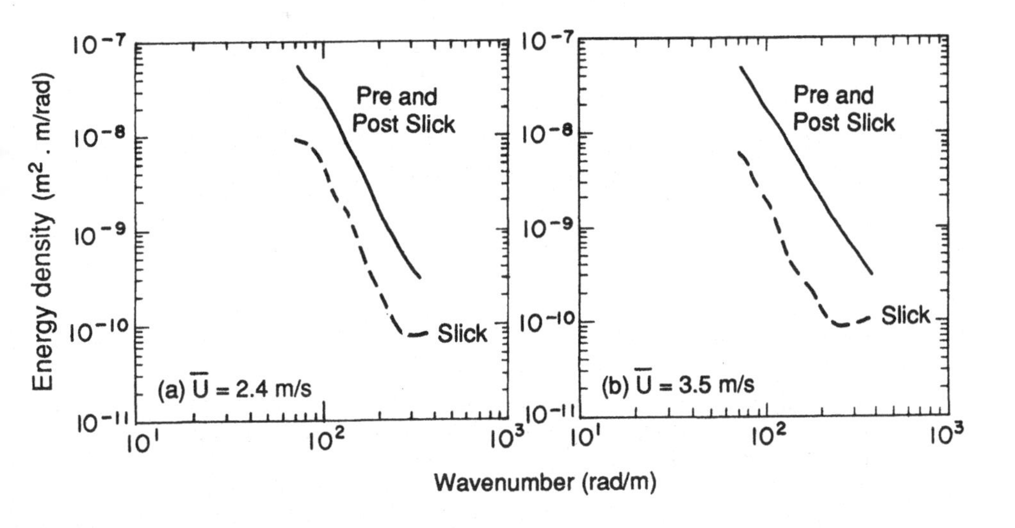

Figure 7. Wave amplitude spectrum, as a function of wavenumber, on spontaneously occurring slicks for two mean wind speeds, $\bar{U}$.

Figure 8 is an illustration of the time evolution of the wavenumber spectra as the slick passed the wave gauge.

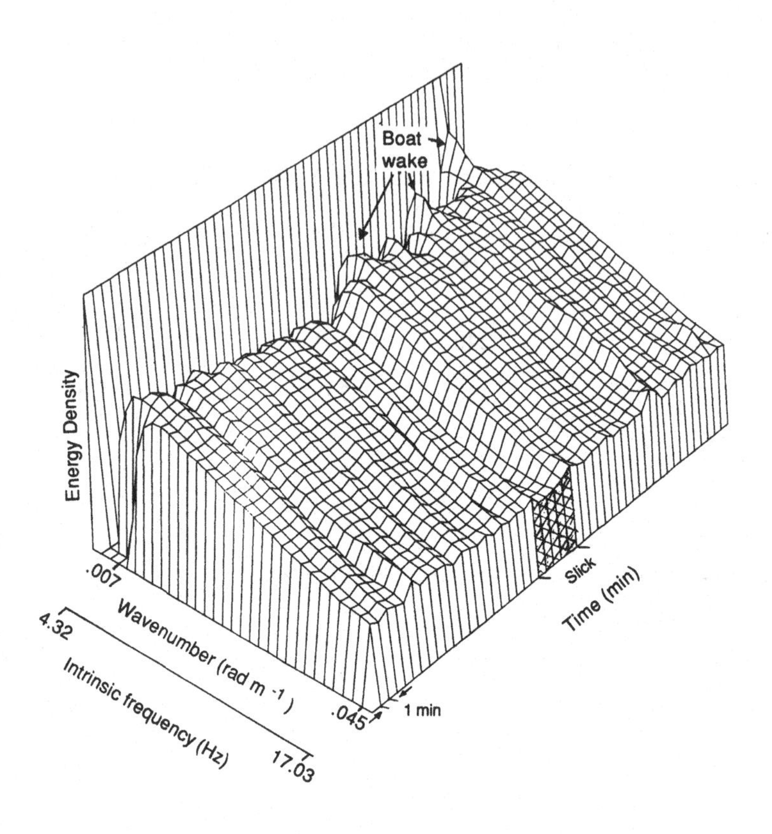

Figure 8. Three dimensional representation of wave amplitude spectra during the passage of a spontaneous slick. The duration of the slick passage at the wave gauge is shaded. Also note the presence of large waves from boat wakes. $\bar{U} = 3.5$ m/s.

Finally, Figure 9 gives the percent change due to the naturally formed slick, of the spectral power density as a function of intrinsic frequency and wavenumber. In this figure, as in Figure 5 slick effects at frequencies > 10 Hz are lost, because the only measurable signal from the slick covered surface is white noise of relatively low amplitude.

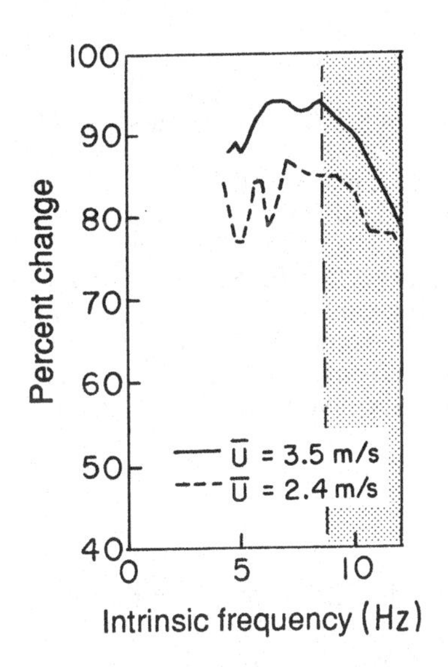

Figure 9. Per cent reduction in spectral power density as a function of intrinsic frequency, for naturally formed slicks. In the cross-hatched region the amplitude of waves on the slick covered surface is less than the resolution of the wave wire.

4. Discussion

The field experiments with model and naturally formed slicks reported here show that effects of compounds modifying surface properties such as surface tension and dilational elasticity are strong for wind speeds as high as 2 to 3 m/s and weaker, but still measurable, at 6 m/s. In contrast, a laboratory study with monomolecular films on a convection tank (Katsaros and Garrett, 1982) showed the effects of the films on evaporation to be removed by even a 1 m/s wind over the surface. Only gravity-capillary and capillary waves (at intrinsic frequencies greater than 2 Hz) were affected by the model slicks, which rarely measured more than 10 m in extent and which were limited to 3 to 5 minute duration at the wave gauge. For the naturally formed slicks, likely to contain biogenic surfactants and some petroleum derivatives the same range of frequencies were affected but wind speed dependence was less consistent. We interpret this as evidence that variation in composition, age, size and other parameters between these slicks strongly influence the wave damping.

A good model slick material must mimic the characteristics of natural materials that are responsible for wave damping. Certainly model compounds should have force-area isotherms and surface dilational elasticity similar to natural materials, while molecular weight distributions, and water solubilities may be secondary parameters that lead to effects of slick age on its damping effectiveness. Our model compound was probably not optimal.

The experiments reported here were exploratory in nature and did not cover the complete range of dominant variables. In the natural laboratory of the Lake Washington field site, a more comprehensive series of experiments, over a range of compound composition, slick size and wind stress could shed light on the energy input to waves of various wavelengths and on the wave-wave interactions that occur with and without slicks.

5. Acknowledgments

The field station at Sand Point, Lake Washington, Seattle, WA, was established with support from the Office of Naval Research Physical Oceanography Program (Grant numbers N00014-84-C-0111 and N00014-85-K-0123 Mod P00003) and the wave analysis techniques were developed with support from the National Aeronautics and Space Administration Ocean Sciences Program (Grant number NAGW-736). Surface tension spreading oils were kindly lent by Dr.Wm Barger, US Naval Research Laboratories, while surfactant measurement support came from the US Environmental Protection Agency. We are grateful to the following persons who assisted in various phases of the project: R.J. Lind and R. Allard assisted with data analysis; J.V. Meadows and K. Dewar produced the manuscript and the figures, respectively; and Dr. Stuart D. Smith made helpful comments on the manuscript..

6. References

Adam, N.K., 1937, 'A Rapid Method for Determining the Lowering of Tension of Exposed Water Surface with some Observations of Surface Tension of the Sea and Inland Waters', *Proc. Royal Soc. (B)*, **122**, 134-139.

Ataktürk, S.S. and K.B. Katsaros, 1987, 'Intrinsic Frequency Spectra of Short Gravity-Capillary Waves Obtained from Temporal Measurements of Wave Height on a Lake', *J. Geophys. Res.*, **92**, 5131-5141.

Ataktürk, S.S., 1988, Characterization of Small Scale Roughness Elements on a Water Surface, Ph.D. Thesis, Dept. Atmos. Sci, Univ. Wash., Seattle,WA, 98195.

Baier, R.E., D.W. Goupil, S. Perlmutter and R.King, 1974, 'Dominant Chemical Composition of Sea-Surface Films, Natural Slicks, and Foams', *J. Recherche Atmos.*, **8**, 571-600.

Carlson, D.J. and L.M. Mayer, 1980, 'Enrichment of Dissolved Phenolic Material in the Surface Microlayer of Coastal Waters', *Nature*, **256**, 482.

Davies, J.T. and R.W. Vose, 1965, 'On the damping of capillary waves by surface films', *Proc. Royal Soc.*, **286**, 218-234.

Garrett, W.D., 1967, 'Damping of Capillary Waves at the Air-sea Interface by Oceanic Surface-active Material', *J. Marine Res.*, **25**, 279-291.

Gottifredi, J.C. and G.J. Jameson, 1968, 'The suppression of wind-generated waves by a surface film', *J. Fluid Mech.*, **32**, 609-618.

Gucinski, H. and D.W. Goupil, 1981, 'Rapid Analysis of Films and Surface Slicks as a Pollutant Monitoring Technique', *Ocean Science and Engineering*, **6**, 358-361.

Gucinski, H., D.W. Goupil and R.E. Baier, 1981, 'The Sampling and Composition of the Surface Microlayer', In *Atmospheric Pollutants in Natural Waters*, S. Eisenreich, ed., Ann Arbor Press, p.165-180.

Herr, F. and J.Williams, 1986, Roll of surfactant films on the interfacial properties of the sea surface ONRL Report C-11-86, Office of Naval Research, London, U.K.

Hino, M., S. Kataoka and D. Kaneko, 1969, 'Experiment of Surface Film Effect on Wave-Wave Generation', *Coastal Engineering in Japan,* **12**, 1-8.

Hühnerfuss, H., W. Alpers, P.A. Lange and W. Walter, 1981, 'Attenuation of wind waves by artificial surface fillms of different chemical structure', *Geophys. Res. Lett.,* **8**, 1184-1186.

Hühnerfuss, H. and W. Alpers, 1983, 'Molecular Aspects of the System Water/Monomolecular Surface Film and the Occurrence of a new Anomalous Dispersion Regime at 1.43 GHz', *J. Phys. Chem.,* **87**, 5251-5258.

Hühnerfuss, H. W. Alpers, A. Cross, W.D. Garrett, W.C. Keller, P.A. Lange, W.J, Plant, F. Schlude and D.L. Schuler, 1983a, 'The Modification of X- and L-Band Radar Signals by Monomolecular Sea Slicks', *J. Geophys. Res.,* **88,** 9817-9822.

Hühnerfuss, H., W. Alpers, W.D. Garrett, P.A. Lange and S. Stolte, 1983b, 'Attenuation of capillary and gravity waves at sea by monomolecular organic surface films', *J. Geophys. Res.,* **88,** 9809-9816.

Hühnerfuss, P.A. Lange and W. Walter, 1985a, 'Relaxation effects in monolayers and their contribution to water wave damping, I, Wave induced phase shifts', *J. Colloid Interface Sci.,* **108**, 430-441.

Hühnerfuss, H., P.A. Lange and W. Walter, 1985b, 'Relation effects in monolayers and their contribution to water wave damping, II, The Marangoni phenomenon and gravity wave attenuation', *J. Colloid Interface Sci.,* **108**, 442-450.

Hühnerfuss, H., W. Walter, P.A. Lange and W. Alpers, 1987, 'Attentuation of Wind Waves by Monomelecular Sea Slicks and the Marangoni Effect', *J. Geophys. Res.,* **92**, 3961-3963.

Katsaros, K.B. and W.D.Garrett, 1982, 'Effects of Organic Surface Films on Evaporation and Thermal Structure of Water in Free and Forced Convection', *Int. J. Heat Mass Transfer* , **25**, 1661-1670.

Liu, H.T., K.B. Katsaros and M.A. Weissman, 1982, 'Dynamic response of thin-wire wave gauges', *J. Geophys. Res.,* **87**, 5686-5698.

Liu, H.T. and J.T. Lin, 1979, Effects of an Oil Slick on Wind Waves, Proc. Oil Spill Conf., Los Angeles, CA, p.665-674.

Lobemeier, P., 1978, 'Entwicklung eines Seegangsmessystems zur Untersuchung des kurzwelligen Seegangs, FWG-Ber. 1978-3', *Forschungsanst. der Bundeswehr fur Wasserschall- und Geophys.,* Kiel, Fed. Rep. of Germany.

Lucassen, J., 1981, 'Effect of Surface-Active Material on the Damping of Gravity Waves: A Reappraisal', *J. Coll. Interf. Sci.,* **85**, 52-58.

Kraus, E.B., 1972, *Atmosphere-Ocean Interaction.*, Clarendon Press, Oxford, 275pp.

Scott, J.C., 1979, *Oil on Troubled Waters: A Bibliography on the Effcts of Surface-Active Films on Surface Wave Motions,* Multi-Science Publ. Co, Ltd. London, UK, 82pp.

Scott, John C., 1972, 'The influence of surface-active contamination on the initiation of wind waves', *J. Fluid Mech.*, **56**, 591-606.

Ulaby, F.T., R.K. Moore and A.K. Fung, 1982, *Microwave Remote Sensing, Active and Passive,Vol II*, Addison-Wesley Publ. Co., Reading, MA., 607 pp.

THE ENERGY BALANCE IN SHORT GRAVITY WAVES

G.J. Komen
KNMI
P.O.Box 201
3730 AE De Bilt
The Netherlands

ABSTRACT. Earlier results on the energy balance in gravity waves (Komen, Hasselmann and Hasselmann, 1985) are discussed. It is shown that these can be used to describe the evolution of short gravity waves, provided wind speeds are low enough. For higher wind speeds equilibrium range ideas are considered.

1. The Energy Transfer Equation

The temporal evolution of surface gravity waves under the influence of wind can be described in terms of the wave spectrum

$$F(k;\underline{x},t) = \frac{1}{(2\pi)^2} \int d\underline{\zeta}\, e^{i\underline{k}\cdot\underline{\zeta}} \langle \eta(\underline{x}+\tfrac{1}{2}\underline{\zeta},t)\,\eta(\underline{x}-\tfrac{1}{2}\underline{\zeta},t)\rangle \tag{1}$$

Here η is the height of the water surface, relative to equilibrium, $\underline{x}$ denotes position and t denotes time. For water waves there is a fixed relationship between frequency and wavenumber (the dispersion relation)

$$\omega^2 = gk \tanh kD \tag{2}$$

with D the local water depth. Therefore, instead of the wavenumber spectrum the frequency directional spectrum is also frequently used. They are related by the transformation Jacobian:

$$F(\underline{k})d\underline{k} = F(f,\Theta)dfd\Theta \tag{3}$$

The total wave variance is equal to the integral over the wave spectrum.

$$\langle\eta^2\rangle = \int d\underline{k}\; F(\underline{k}) \tag{4}$$

The significant wave height is four times the square root of the variance.

For deep water, the energy transfer equation reads,

G. J. Komen and W. A. Oost (eds.), Radar Scattering from Modulated Wind Waves, 75–79.

$$\frac{\partial F}{\partial t} + \underline{c}\,\frac{\partial F}{\partial x} = S_{in} + S_{nl} + S_{ds} \qquad (5)$$

The left hand side gives the total rate of change for each frequency-directional wave component when you move with that wave at its own group velocity $\underline{c}$.

The so-called source terms on the right hand side describe the physical processes that govern the evolution of the spectrum. S_{in} represents the wind input, it is linear in the spectrum, and the coefficient depends explicitly on the wind. S_{nl} represents nonlinear resonant four-wave interaction, and is given in terms of a Boltzmann integral. Finally, S_{ds} expresses dissipation processes. For a full discussion of these source terms in this context we refer to Komen, Hasselmann and Hasselmann (1985), hereafter KHH.

KHH solved equation 5 numerically to study the dynamics of wave generation in detail. For computational reasons they only considered the range of frequencies from 0.04 Hz to 2.5 times the peak frequency. Above this cutoff they parametrized the spectral tail and they did not consider the detailed dynamics. An example of their results is given in figure 1, where the balance between the different source terms in the equilibrium limit of fetch limited growth is given.

A striking feature of figure 1 is its complexity: the balance is not a simple one. In particular one does not have a simple power law for the spectral shape; also the angular dependence is rather non-trivial. All of these features can be qualitatively understood in terms of the behaviour of the source terms.

2. Relevance for Short Gravity Waves

The computations outlined above are valid only in a range around the peak of the spectrum. For strong wind and developed sea the corresponding wave components will be fairly long, so that the discussion of section 1 appears not to be relevant for our understanding of the very short gravity waves with a wave length of the order of a decimeter. Table 1 gives an idea of the length scales involved. For two different wind speeds (10 meter winds) and for two different stages of development (defined as phase speed at the peak frequency divided by the 10 m wind) the wave length at the peak and the wave length at 2.5 times the peak frequency are given. At the lowest wind speed considered (2.5 m/s or approximately 5 knots) and for fairly young wavelets the dynamic range extends to 16 cm. Of course, at higher wind speeds these short waves fall outside the dynamic range.

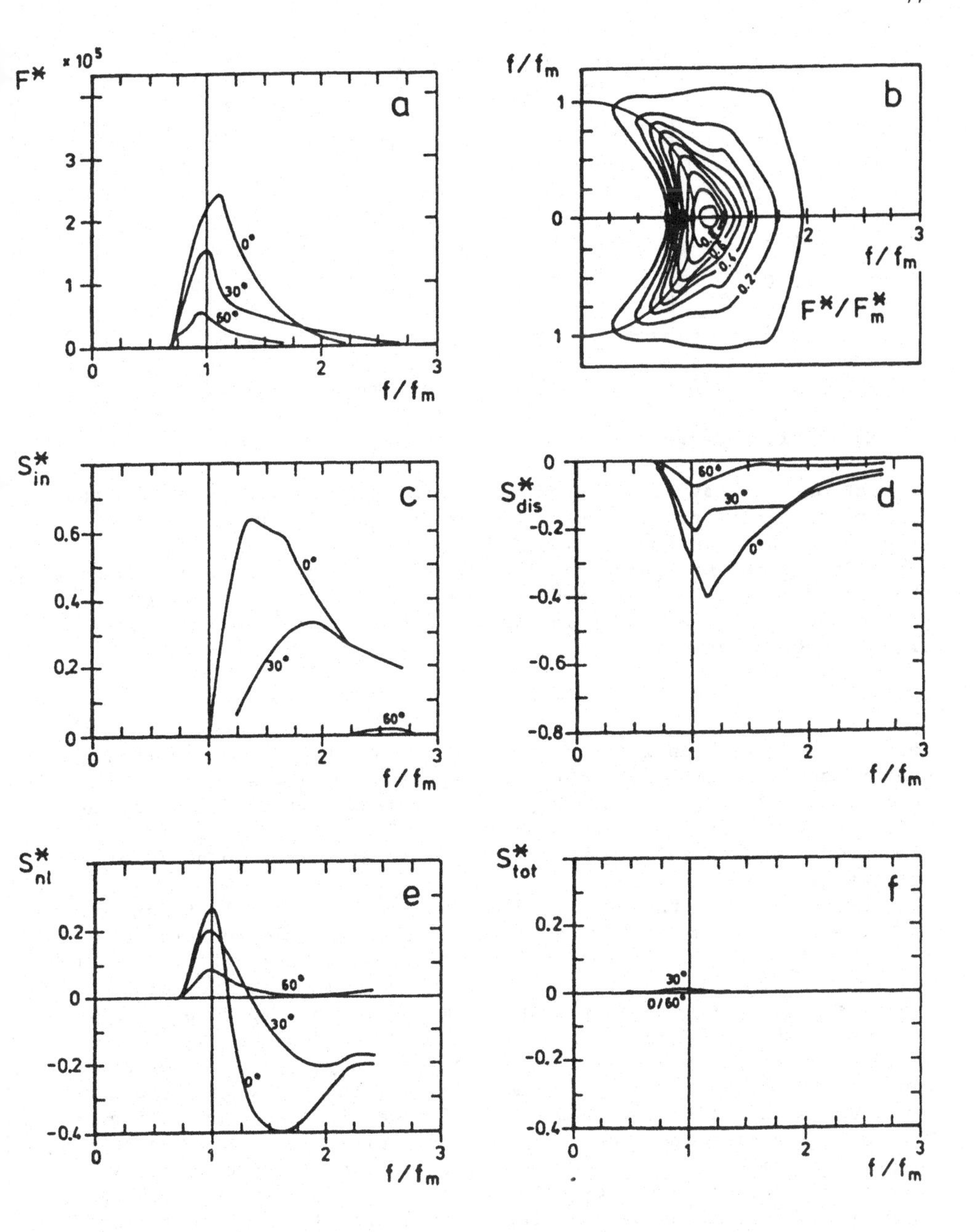

Figure 1. Illustration of the balance between different source terms in the energy transfer equation (taken form KHH). The top boxes give the wave spectrum for different directions relative to the wind. The lower boxes give the separate source terms and their sum. All functions were made nondimensional with the help of friction velocity.

	c_p/U_{10}=0.75			c_p/U_{10}=0.5		
U_{10} (m/s)	c	L_p	L_c	c	L_p	L_c
2.5	1.9	2.2	0.36	1.25	1.00	0.16
5.0	3.75	9.0	1.44	2.5	4.0	0.54

TABLE 1 Length scales of the dynamic range (in meters).

3. Equilibrium Range

In section 1 we have considered the dynamics of growing waves in the energy containing range around the peak of the spectrum. What can be said about the high frequency tail?

Phillips (1958) has suggested that for short waves at high frequency the spectral level is mainly determined by breaking; waves grow until they break, and this saturation level is universal. On dimensional arguments one then has

$$F(k) = \alpha\, k^{-4}, \tag{6}$$

corresponding with an f^{-5} spectrum. It is remarkable that in this context one sometimes refers to Jonswap as experimental evidence for the validity of (6). Indeed, Jonswap found evidence for an f^{-5} tail, but this was in the frequency range between 1.2 and 2 times the peak frequency. As we have seen above asymptotic arguments do not apply in this range.

New evidence seems to indicate that (6) is incorrect. It turned out that the constant α depended on fetch. Also, it would be hard to reconcile the idea of measuring wind by means of backscatter from short waves if there were a universal spectral tail. In fact Toba et al have presented evidence that an f^{-4} spectrum is more appropriate.

Kitaigorodskii et al have brought forward the idea that simple scaling arguments might be used to explain the frequency dependence of ocean waves in the high frequency regime. They use analogy with turbulence, where there is input of energy at low frequencies and dissipation at the high end. Transfer from low to high frequencies (the so-called energy cascade) proceeds through nonlinear interactions. This picture is an attractive one. However, it cannot be correct, because in ocean waves there is no separation between input scales and dissipation scales. This led Phillips (1985) to extend Kitaigorodskii's ideas. He assumed that all source terms are important and that they are of the same magnitude. Introducing B as k^4F , he assumes

$$S_{diss} = gk^{-4}f(B(k))$$

$$S_{in} = gk^{-4}(u_*/c)^2B(k)$$

$$S_{nl} = gk^{-4}B^3(k)$$

It is essential to be at high frequency ($f \gg f_p$) for the locality of the nonlinear transfer to be valid. (In reality it is quasi-local, but if you have a power tail, spectral levels around a particular wavenumber will be related to the spectral level at that wave number). Assuming $S_{in} = S_{nl}$ you then have

$$F(k) = u_* g^{-1/2} k^{-7/2} \qquad (7)$$

which corresponds to an f^{-4} tail.

4. Conclusion

We have shown that for low wind speeds the results of KHH can be used to discuss the dynamics of short gravity waves. For higher wind speeds the scaling arguments of Phillips might be applicable.

We recommend an extension of the work of KHH to a much larger frequency range, so that the assumptions of Phillips can be explicitly verified. In such calculations capillary effects and the associated three wave interaction could be included.

References

Kitaigorodskii, S.A., V.P. Krasitskii and M.M. Zaslavskii (1975) 'On Phillips' theory of equilibrium range in the spectra of wind-generated gravity waves.' J. Phys. Oceanogr. **5**, 410.

Komen G.J., S. Hasselmann and K. Hasselmann (1984) 'On the existence of a fully developed wind-sea spectrum.' J. Phys. Oceanogr. **14**, 1271.

Phillips, O.M. (1958) 'The equilibrium range in the spectrum of wind-generated gravity waves.' J. Fluid Mech. **66**, 625.

Phillips, O.M. (1985) 'Spectral and statistical properties of the equilibrium range in wind-generated gravity waves.' J. Fluid Mech. **156**, 505-531.

Toba, Y (1973) 'Local balance in the air-sea boundary process III On the spectrum of wind waves.' J. Oceanogr. Soc. Japan **29**, 209.

DERIVATION OF PHILLIPS α-PARAMETER FROM TURBULENT DIFFUSION AS A DAMPING MECHANISM

W. Rosenthal
GKSS Research Centre Geesthacht, Institute of Physics
P. O. Box 1160
2054 Geesthacht, Federal Republic of Germany

ABSTRACT. A damping mechanism nonlinear in energy density is proposed to be responsible for wave dissipation in the open ocean. It is then possible to calculate an order of magnitude to Phillips α-parameter, defining the energy content in the saturation range of the spectrum.

1. INTRODUCTION

It has been argued from dimensional consideration first by Phillips (1958) that the frequency dependence of the one-dimensional surface wave spectrum is $\alpha\, g^2 (2\,\Pi)^{-4}\, f^{-5}$ in a spectral region above the peak frequency and that this shape is rather independent on wind speed. This region is called the saturation range. The appropriate spectrum in k space is βk^{-4} with a dimensionless parameter β that is half the value of α. The lower frequency part of a wind wave spectrum is modelled by multiplying a shape factor such that the spectrum exhibits a sharp peak and declines rapidly towards lower frequencies. Recently it was realized that at least certain parts of the spectrum are more to a f^{-4} behaviour than f^{-5}. Stolte (this volume) found for frequencies higher than 8 Hz a f^{-2} slope. It may, however, be mentioned that in the presence of longer waves the frequency dependence is normally measured in terms of frequency of encounter. A careful analysis must be applied to transform spectra from frequency of encounter to intrinsic frequency (Richter, Rosenthal 1981). In the following we want to derive a more physical explanation for the shape of the saturation range that builds on generation and dissipation processes for each fixed frequency-direction band.

2. THE NONLINEAR DISSIPATION

The energy balance equation, leading to the spectral shape is normally written as

$$\frac{\partial E}{\partial t} + \underline{V}_G \cdot \frac{\partial E}{\partial \underline{x}} = S = S_{in} + S_{nl} + S_{dis} \qquad (1)$$

G. J. Komen and W. A. Oost (eds.), Radar Scattering from Modulated Wind Waves, 81–88.

where E and S are functions of frequency and angle relative to wind direction (see G. Komen's review in this proceedings). For a homogeneous and stationary situation the equation simplifies to

$$S = 0 \tag{2}$$

In publications by Plant (1986) and Donelan, Pierson (1987), S has been developed into a Taylor series with respect to energy density up to the quadratic term for a given frequency-direction band

$$S = \frac{\partial S}{\partial E(f)}\Big|_f \; E(f) + \frac{\partial^2 S}{\partial E(f)^2}\Big|_f \; E^2(f) = a_1 E(f) + a_2 E^2(f) \tag{3}$$

From equation (2) this will yield the spectral shape

$$E(f) = -\frac{a_1}{a_2} \tag{4}$$

With ad-hoc assumptions on wave number dependence of a_1, a_2 the observed or assumed spectral shape has been obtained by the authors mentioned above. It is however open to discussion which physical mechanism produces the $E^3(f)$ dependence of the source function. In the following section we want to investigate the process of vertical momentum diffusion (momentum exchange by turbulent eddy viscosity) for producing a term quadratic in E(f) in equation (3).

3. OUTLINE OF PROCEDURE

We start from the mass balance and momentum balance equations

$$\frac{\partial \rho_w U_i}{\partial t} = -\frac{\partial}{\partial x_j}\tau_{ij}$$

$$i,j = 1,\ 2,\ 3 \tag{5}$$

$$\frac{\partial \rho_w}{\partial t} = -\frac{\partial \rho_w U_j}{\partial x_j}$$

$$\tau_{i,j} = \rho_w U_i U_j + \delta_{ij}(p + \Phi) + A\frac{\partial U_i}{\partial x_j}$$

$$\Phi = \rho_w g x_3$$

A = diffusion coefficient (eddy exchange coefficient)
(variable names are standard and explained e.g. in Dolata, Rosenthal)

We refer to Longuet-Higgins (1953) or to Phillips text book (1976) on the generation of vorticity and a downward momentum transport by means of the last term in $\tau_{i,j}$ when A is taken as the molecular viscosity.

The molecular viscosity is of the order of $1.0 \cdot 10^{-6}\ m^2s^{-1}$ whereas for the eddy exchange coefficient measurements between 100 and 10 000 times $10^{-4}\ m^2s^{-1}$ are found in the literature (Hoeber, 1972).

These large values for A, treated in the same manner as molecular viscosity would lead to large exponential decay rates β of surface waves, since in deep water for wave number k:

$$\beta = 2Ak^2 \tag{6}$$

Nevertheless there is literature (Weber, 1983) in which these large decay rates are postulated.

The observed decay rates of surface waves are much smaller than those defined by (6). An explicit criticism of the derivation of (6) by taking over the deductions from molecular diffusion will be given elsewhere. We mention only, that the development of a viscous surface layer is necessary for the extraction of momentum from surface waves, while outside the viscous layer the contribution

$$\frac{\partial}{\partial x_j}\left(A\,\frac{\partial U_i}{\partial x_j}\right)$$

in the momentum balance equation (5) vanishes for harmonic surface waves if A can be treated as constant. To explain the fact that (6) gives too large decay rates we may assume that no surface layer is formed due to the eddy viscosity term, e.g. there is no near surface layer in which this term dominates the other constituents of the momentum balance equation.

We consider the Stokes solution for the velocity potential Φ with $\underline{u} = \underline{\nabla}\Phi$

$$\Phi = v_0 a^2 t + v_1 a \cosh k \cdot (h + z) \sin\Theta$$
$$+ v_2 a^2 \cosh 2k(h + z) \sin 2\Theta$$

$$v_0 = \frac{gk}{2\sinh 2kh} \qquad v_1 = \frac{w_0}{k \sinh kh} \qquad v_2 = \frac{3}{8}\,\frac{w_0}{\sinh^4 k \cdot h}$$

$$\frac{w^2}{g K \tanh kh} = 1 + k^2a^2\left(\frac{9 \tanh^4 kh - 10 \tanh^2 kh + 9}{8 \tanh^4 k h}\right) \tag{7}$$

(Witham)

Since all terms are solutions of Laplace equation

$$\frac{\partial}{\partial x_i}\frac{\partial}{\partial x_i}\Phi = 0 \tag{8}$$

there is no contribution in the equation of motion from the diffusion term if we assume a constant A. This changes if we consider the velocity, augmented by the Stokes drift.

First order plus Stokes drift

$$\eta = a \cos(\underline{k}\ \underline{x} - \omega t) \qquad \text{surface elongation}$$

$$U_\alpha = \frac{k_\alpha}{k}\ \sigma\ a \cos(\underline{k}\ \underline{x} - \omega t)\ \frac{\cosh k(x_3 + h)}{\sinh kh} \qquad \alpha = 1, 2$$

$$+ g \frac{k_\alpha}{\sigma}\ k\ a^3\ \frac{\cosh 2k(z + h)}{\sinh 2\ k\ h}$$

$$U_3 = \sigma\ a \sin(\underline{k}\ \underline{x} - \omega t)\ \frac{\sinh k(x_3 + h)}{\sinh k\ h} \tag{9}$$

Energy-change in time is given by:

$$\frac{\partial E}{\partial t} = \int dxdydz\ \{\rho\ U_i \frac{\partial U_i}{\partial t}\} + \int dxdy\ \rho\ g\ \eta \frac{\partial \eta}{\partial t} \tag{10}$$

In the expression

$\frac{\partial}{\partial x_j} \tau_{ij}$ we consider the expression for turbulent diffusion

$$A \frac{\partial^2}{\partial x_j{}^2}\ U = A\ k^2 \frac{g}{\sigma} k_\alpha\ \ k\ a^2\ \frac{\cosh 2k(z + h)}{\sinh 2\ k\ h} \tag{11}$$

where only the Stokes drift gives a non-vanishing contribution.

Multiplication of $\frac{\partial\ U_\alpha}{\partial t}$ with U_α and integration in x and y gives zero, and therefore

$$\int dxdydz U_\alpha \frac{\partial U_\alpha}{\partial t} = 0$$

and the kinetic energy stays constant if the diffusion term is neglected. Integrating the diffusion term over space gives

$$I = \int\ A\ U_\alpha \frac{\partial^2}{\partial x_j{}^2}\ U_\alpha\ dz\ dx\ dy$$

$$I = \Omega\ \frac{A}{4}\ \frac{g^2}{\sigma^2}\ k^5\ a^4 \{\coth 2\ k\ h + \frac{k\ h}{\sinh^2 k\ h}\} \tag{12}$$

Ω = horizontal normalization volume

Energy loss per surface unit due to the diffusion is therefore,

$$S_{dif} = \frac{\partial E}{\partial t} = - \frac{I}{\Omega} = - \frac{A}{4} \frac{g^2}{\sigma^2} k^5 a^4 \{\coth 2 k h + \frac{k h}{\sinh^2 k h}\} \quad (13)$$

$$\text{in deep water} = - \frac{A}{4} g k^4 a^4$$

$$\text{else} = - \frac{A}{4} g k^4 a^4 \{ \frac{\coth 2 k h}{\tanh k h} + \frac{k h}{\sinh^2 k h \tanh k h} \} \quad (14)$$

We use the equation (2), that means $\frac{\partial E}{\partial t}$ and $V_{\underline{G}} \frac{\partial E}{\partial \underline{x}}$ is neglected in the balance equation for the energy. The generation of wave energy is usually modelled by

$$S_{in} = \beta f (\frac{U}{c} - 1) g \frac{a^2}{2}$$ for $U > c$ being the surface wind

To take into account the nonlinear interaction we add a term linear in energy, e.g. linear in a^2:

$$S_{nl} = y \cdot g \frac{a^2}{2}$$

where y has the dimension of a frequency.

Inserting in (2) the source function $S = S_{in} + S_{nl} + S_{dif}$ we get for deep water

$$a^2(k) = \frac{\frac{1}{2} [\beta f(\frac{U}{c} - 1) + y]}{[\frac{A}{4} k^4]} \quad (15)$$

In general we have

$$a^2 (k) = \frac{2 [\beta f (\frac{U}{c} - 1) + y]}{[A k^4 \{\frac{\coth 2 k h}{\tanh k h} + \frac{k h}{\sinh^2 k h \tanh k h}\}]} \quad (16)$$

Since the actual energy per m^2 is calculated by multiplying with ρ_{water} (mass per m^3) we have to multiply with 1 m^2 to get the dimensions correct. The result for spectral energy density is then

$$a^2(k) = b k^{-4}$$ with dimensionless b

From measurements we know the value of Phillips parameter α. It can be shown that

$$b = \frac{\alpha}{2} \tag{17}$$

From the above calculations

$$\alpha = \alpha_D = 4 \cdot [\beta\, f(\frac{U}{c} - 1) + y] \,/\, A \cdot 1\ m^2 \qquad \text{for deep water}$$

$$\alpha = \alpha/[\frac{\coth 2\,k\,h}{\tanh k\,h} + \frac{k\,h}{\sinh^2 k\,h \tanh k\,h}] \qquad \text{for finite depth } h \tag{18}$$

$$\beta\, f\,(\frac{U}{c} - 1) + y \quad = \quad \text{Order of } 10^{-3}\ s^{-1}$$

With typical orders of magnitude for the eddy viscosity A

$$100 \cdot 10^{-4}\ m^2 s^{-1} < A < 10000 \cdot 10^{-4}\ m^2 s^{-1} \tag{19}$$

we arrive at orders of magnitude for α:

$$0.004 < \alpha < 0.4 \tag{20}$$

These orders of magnitude may be compared with typical values for α in figure 1 from the JONSWAP report (Hasselmann et al.).

4. CONCLUSIONS

It has been shown that the eddy exchange term in the momentum balance of surface waves can be the origin of dissipation that is nonlinear in terms of wave momentum. This dissipation can explain the high frequency shape of the spectrum. It can, at least qualitatively, explain the variation of the Phillips α-parameter. In fig. 1 for instance the short fetch values of α are measured in wave banks where we can assume a rather small eddy diffusion coefficient A. The decreasing α values with increasing distance from shore indicate an increasing value of A, which could be expected from the increased turbulence created for instance by increased number of breaking events.

5. REFERENCES

Phillips, O. M., 1958, J. Fluid Mech. 4 (1958a) 426.

Richter, K., Rosenthal, W., 1983: Energy Distribution of Waves above 1 Hz on Long Wind Waves, in: "Wave dynamics and Radio Probing of the Ocean Surface", edited by O. M. Phillips and K. Hasselmann, Plenum Press, 1986.

Donelan, M. A., Pierson, W. J. P., 1987: Radar Scattering and Equilibrium Ranges in Wind-Generated Waves with Application to Scatterometry, JGR 92, Nr. C5, 4971-5029.

Plant, W. J. P., 1986: A Two-Scale Model of Short Wind-Generated Waves and Scatterometry, JGR 91, Nr. C9, 10735-10749.

Longuet-Higgins, M. S., 1953: Mass Transport in Water Waves, Phil Trans A, 245, 535-581.

Phillips, O. M., 1976: The Dynamics of the Upper Ocean, Cambridge University Press.

Hoeber, H., 1972: Eddy thermal conductivity in the upper 12 m of the tropical Atlantic, JPO 2, 303-304.

Weber, J. E., 1983: Attenuated wave induced drift in a viscous rotating ocean, Journ. Fluid. Mech., 137, 115-129.

Dolata, L. F., Rosenthal, W., 1984: Wave Setup and Wave-Induced Currents in Coastal Zones, JGR 89 C2, 1973-1982.

Witham, 1974: Linear and Nonlinear Waves, Wiley and Sons.

Hasselmann, K., Barnett, T. P., Bouws, E., Carlson, H., Cartwright, D. E., Enke, K., Ewing, J. A., Gienapp, H., Hasselmann, D. E., Kruseman, P., Meerburg, A., Müller, P., Olbers, D. J., Richter, K., Sell, W., Walden, H., 1973: Dtsch. Hydrogr. Z. A(8°), No. 12.

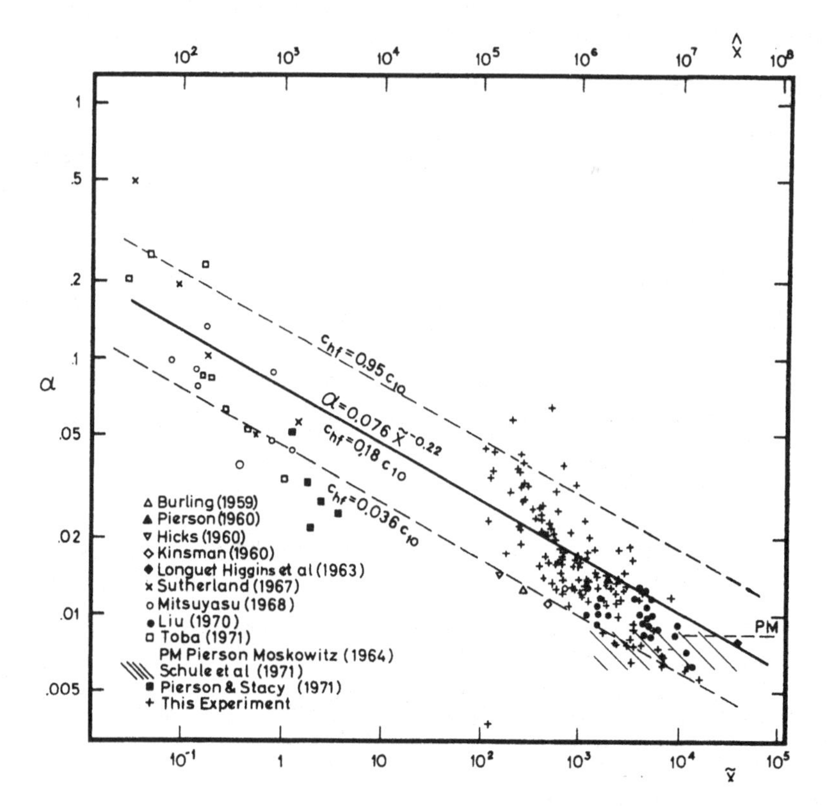

Figure 1. Variation of α with dimensionless fetch $\tilde{x} = \frac{gx}{U_{10}^2}$ (Hasselmann et al.)

TEMPORAL AND SPATIAL VARIABILITY OF THE WIND STRESS VECTOR

Gerald L. Geernaert
U.S. Naval Research Laboratory
Washington, DC 20375-5000
USA

ABSTRACT. In this paper, we present observations and some theoretically-based arguments to gain insight on periodicities in the wind stress vector at the ocean surface. These periodicities, illustrated by measurements collected during a number of recent field experiments, indicate that the inertial time scale, planetary boundary layer (PBL) rolls, and dominant wavelengths of surface gravity waves and swell become important in interpreting stress estimates. For long averaging, the surface wave field is an important feed-back component in predicting the magnitude of the stress vector. However, for problems concerned with shorter time or length scales, the variation of the stress vector according to phase of the long wave must be considered; such problems include the long wave phase dependent growth rates of short waves and wave/current interactions.

I. INTRODUCTION

When wind blows over a water surface, the resulting vertical wind shear supports a downward flux of momentum, a process more commonly referred to as wind drag or wind stress. Momentum will enter the air-sea interface producing a mean drift current, wave instabilities, and turbulence: With time duration, the wave instabilities will grow to produce a spectrum of gravity and capillary waves, and if the wind speed is high enough (>7 m/sec), the waves may become locally unstable and break. The breaking process in turn aids the momentum (and heat) transfer through turbulent entrainment across the interface. Since ocean waves have surface-projected orbital velocity vectors which are phase dependent, wave-induced pressure fluctuations in the wind field immediately above the surface may act to modulate the local wind shear and consequently the net drag. The physical problem of evaluating the wind stress on scales that are small compared to the longest gravity waves is therefore a complicated problem of feedbacks and nonlinearities.

The common approach to understanding the relationship between the wave state and wind stress has been to ignore the microscale problem of

G. J. Komen and W. A. Oost (eds.), Radar Scattering from Modulated Wind Waves, 89–104.

momentum transfer and to assume that the surface is homogeneous. This assumption allows a simple roughness scale (normally the roughness length z_0) to be employed to describe the drag that the spatially-averaged wave field exerts on the wind. Since wave spectra carry a systematic dependence on fetch and depth (refer to Hasselmann, et al., 1975), modeling of wave-induced variability in the average statistics of stress is often based on the variability of upwind fetch and/or water column depth between measurement sites.

While the physical problem of wind drag must consider a large number of surface and turbulence parameters, the practical approach is based on incorporating bulk meteorological and oceanographic variables mainly from ships and buoys. This bulk aerodynamic technique, therefore, relies on the statistical analysis of data collected over long temporal or spatial time series, where the data include only easily measured quantities such as wind speed, air and water temperatures, and humidity. For near neutral stratifications, the wind stress scales very closely to the square of the wind speed; if the stratification deviates from neutrality, significant adjustments must be made to the initial neutral stress estimate. Because the stratification function, wind speed, and temperature profile are height dependent, experimentalists generally normalize their analyses, and modelers base their data on measurements representative of the 10 m elevation above the surface.

Wind stress calculations are widely performed in a number of scientific problems. Numerical prediction of oceanic and atmospheric circulations require high quality information on the absolute magnitude and spatial variability of the surface stress vector. Remote sensing algorithms (especially the altimeter, scatterometer, and SSM/I) are based on surface signatures that in turn are dependent in large part on the local stress. For both numerical prediction and space-based remote sensing campaigns, mesoscale and/or macroscale estimates of the stress are generally adequate. For these estimates, the mean statistics of a horizontally homogeneous surface will suffice while little information is required on the microscale processes associated with wave/wave interactions.

However, when one must investigate problems associated with air modification, boundary layer cloud dynamics, coastal processes, sea surface temperature fronts, storms, and radar ducts, the horizontal homogeneity criterion becomes an over-simplifying assumption and more attention must, therefore, be paid to the actual mechanism of momentum transfer to the wavy air-sea interface. In this case, surface wave slopes (including both long and short waves) and surface tension must be considered.

In this paper, we will focus on variability of the stress vector due to boundary layer and inertial constraints on one hand, and to surface wave patterns on the other. In the next section, the traditional approach to parameterizing the wind drag will be presented. Periodicities due to boundary layer rolls and larger scale motions will be discussed in Section 3 with illustrations using North Sea and coastal California data. In Section 4, periodicities due to wave patterns will be described using MARSEN data, while in Section 5, we will summarize the implication of these results within the context of wave/current and wave/wave interactions within a mesoscale region.

2. TRADITIONAL RELATIONS

The wind stress vector, τ, is classically defined according to:

$$\tau/\rho = - \langle u'w' \rangle i - \langle v'w' \rangle j \tag{1}$$

where u', v', and w', are fluctuations in the longitudinal, lateral, and vertical directions, respectively; ρ is the air density; and the averaging time is chosen such that the mean vertical velocity, w, approaches zero. Although the averaging time depends on both the measurement height and wind speed (Wyngaard, 1973), most investigators choose 30-60 minutes as the averaging time for measurements collected over the ocean from fixed platforms. The implication of such a long averaging time is that the spatially integrated averaging distance is two or three orders of magnitude longer than the longest locally generated gravity wavelength.

Using (1), one may define an angle θ between the stress and wind vectors to be:

$$\theta = \langle v'w' \rangle / \langle u'w' \rangle \tag{2}$$

Geernaert (1988) determined from North Sea measurements that θ exhibits a statistical dependence on the heat flux, i.e.,

$$\theta^{o} = 16 \, z \, \langle w'T_v' \rangle \tag{3}$$

where z is the measurement height and T_v' represents the fluctuating virtual temperature, and the units in (3) are MKS. The quantity $\langle w'T_v' \rangle$ is the heat flux. For z=10m, $|\theta|$ is generally less than 10°.

While remaining within the neutrally-stratified well-mixed surface layer (typical depths on the order of 50 m), the profile of mean wind speed above the ocean surface may be determined by assuming that the turbulent mixing scale is proportional to the height, z. Referring to Businger (1973), one obtains:

$$U_z - U_0 = |\tau/\rho|^{1/2} k^{-1} \ln(\{z+z_o\}/z_o) \tag{4}$$

where U_z represents the wind speed at height z while U_0 is the projected surface current speed. The von Karman constant has an experimentally determined value of 0.4 (Zhang, 1988), while the roughness length, z_o, is assumed to depend on sea state. Since the quantity $|\tau/\rho|$ has units of m/sec squared, one usually introduces the friction velocity u*, such that (4) is rewritten as:

$$U_z - U_0 = u_*/k \ln (z/z_o) \tag{5}$$

where we have made use of the fact that $z >> z_o$. To account for diabatic conditions, equation (5) is often written in the form:

$$U_z - U_0 = u*/k\ (\ln\{z/z_o\} - \psi_M) \qquad (6)$$

The function ψ_M in (6) is born out of a theoretical development for unstable stratifications and a Taylor's series approximation for stable stratifications (after Businger, et al., 1971). We now have:

$$\psi_M = \begin{cases} 2\ \ln[(1+\phi_M^{-1})/2] + \ln(1+\phi_M^{-2})/2 - 2\ \tan^{-1}\phi_M^{-1} + \pi/2; & \text{unstable flow} \\ -\ 5\ z/L;\ \text{stable flow} & \end{cases} \qquad (7)$$

where $\phi_M = (1 - 16\ z/L)^{1/4}$.

The Monin-Obukhov length, L, found in (7) is usually normalized with height so as to become a dimensionless parameter (see Busch, 1973), i.e.,

$$z/L = -gkz\langle w'T_v'\rangle/(T_v\ u*^3) \qquad (8)$$

where g is gravity. For positive upward heat flux (unstable flow), z/L is negative while for stable stratifications, z/L is positive. Operationally, it is more practical to estimate the heat flux using bulk methods; based on Smith (1980), we can write

$$\langle w'T_v'\rangle = C_H U_{10}(T_0 - T_{10}) \qquad (9)$$

where subscripts "0" and "10" represent the surface and ten-meter height respectively, and values of the Stanton number, C_H, are 1.08 X 10^{-3} for unstable flow and 0.86 X 10^{-3} for stable stratifications. Note that (9) is based here on the assumption that humidity effects are much smaller than temperature effects in the total heat flux. (This assumption is valid for surface temperatures that are generally below 24°C.)

Rearranging (6), one easily obtains:

$$u*^2 = C_D\ (U_z - U_0)^2 \qquad (10)$$

where the drag coefficient, C_D, is:

$$C_D = [k/(\ln\{z/z_o\} - \psi_M)]^2 \qquad (11)$$

One can clearly see from (11) that the drag coefficient may be physically interpreted as:

$$C_D(z, z_o, z/L) = C_D\ (\text{height, wave state, stratification}) \qquad (12)$$

Equation (12) could imply that a normalization of the drag coefficient for a height of ten-meters, neutral stratification, and "steady-state" sea state dependence on wind speed could lead to a constant value. An additional function that has been left out of (12) is the nature of atmospheric turbulence, i.e., periodic flows, intermittency, rolls, and non-homogeneous dynamics. Due to our inability to currently describe this additional function, we will only state that the drag coefficient can to first order be normalized by parameterizing with easily measured features of the surface layer, i.e., the measurement height and stratification. The sea state and natural turbulence are usually blamed for producing the remaining scatter within data sets. Most experimentalists report the ten-meter height, neutral-stratification counterpart to C_D, i.e., C_{DN}, so as to compare experimentally collected data sets of differing average stratifications. (Data sets collected at one site during the summer may have significantly different mean stratifications compared to the autumn or winter.) By writing equation (6) into its neutral form and comparing to its diabatic counterpart (see Geernaert and Katsaros, 1986), one obtains an approximate solution, i.e.,

$$C_D = [C_{DN}^{-1/2} - \psi_M/k]^{-2} \tag{13}$$

It must be pointed out that both C_D and C_{DN} represent both long time duration and large spatial distances, which significantly exceed scales associated with the surface wave field. It is on this premise that wave effects on the drag coefficient are based on surface statistics that are spatially homogeneous, and surface features are built into a single roughness parameter, i.e., z_o. For an illustration of the distribution of the measured C_{DN} with wind speed (for a ten-meter measurement height), refer to data collected from the North Sea Platform in Figure 1. While the data in Figure 1 correspond to a constant measurement height and neutral stratifications, the obvious remaining scatter is presumed to be associated with variabilities in turbulence and wave state.

Theorists more-or-less agree that the friction drag induced by short waves supports the bulk of the stress, particularly since the shorter waves carry the largest contribution to the wave slope spectrum. Steeper short waves have been argued to be associated with higher surface stresses (Plant, 1982; Melville, 1977; Geernaert, et al., 1986; Phillips, 1980, etc.). Even though short waves vary along the phase of long waves and have energy densities that depend on the background currents, modeling the wave effect has been approached by considering bulk easily measured wave parameters. Two bulk parameters have been considered: the wave age (c_o/U), used, for example, by Donelan (1982); and the parameter $c_o/u*$, used, for example, by Geernaert, et al. (1986, 1987) and Hsu (1975). For both of these bulk wave parameters, smaller values are associated with steeper short waves and higher drag coefficients. Geernaert, et al. (1988) and Barger, et al. (1970) have additionally shown that in the presence of low surface tension and/or slicks, the drag coefficient (and short wave steepness) are dramatically reduced.

Theoretical approaches designed to tackle the wind stress/wave state problem include those of Donelan (1982), Kitaigorodskii (1973), Huang, et al. (1986), and Hsu (1974). In the latter two models, the basis stems from the Charnock (1955) relation, which states that the roughness length scales with $u*^2/g$. All four models require some information on the parameterized surface wave spectrum; validation of each model depends considerably on tuning and calibration with fetch, depth, and/or time duration effects which bias the characteristic wave conditions at the respective sites of interest.

Experimental evidence indicates that growing waves, short fetch, shallow water, and shoaling conditions, all are associated with larger than "average" drag coefficients in the absence of background swell. In the presence of swell propagating against a wind-driven sea, drag coefficients have similarly been found to be larger than the "average". A wind-driven sea in the presence of swell propagating in the same direction is conversely associated with "smaller" than average drag coefficients. Donelan (1986) has found from wave tank experiments that the short wave energy is reduced in the presence of swell propagating in the same direction as the wind, a finding which is consistent with the field evidence. In the case of background tidal currents, Geernaert (1983) determined that the drag was higher for tidal currents opposing the flow than for currents with the flow; this finding is consistent with a study conducted by Kitaigorodskii and Razumov (1978) on the interaction of currents with wind waves.

In the next section, we will focus on spatial variability with an emphasis on the boundary layer depth scale.

3. LARGE SCALE PERIODICITIES

If one takes a downward looking view from a space-borne satellite, dominant ocean surface length scales in interfacial patterns may readily be seen (especially during light wind speeds). In slightly unstable regions, cumulus clouds will appear to be separated by two or three kilometers; and if one is on the ground, rainbands may appear periodically, i.e., on the order of one rainy spell every 1 to 3 hours. Studies have shown that periodicities in mesoscale divergence patterns occur with roughly 1.5 times the planetary boundary layer (PBL) depth. Over the ocean, this corresponds in the mid-latitudes to horizontal length scales on the order of 2 to 5 kilometers, with the larger numbers associated with higher wind speeds. These length scales have similarly been found in SAR imagery of ocean surface cross-section intensities, and are, therefore, assumed here to be due in part on PBL rolls.

The accuracy of drag coefficient estimates for use in meso- and macro-scale air and ocean models requires that both a statistically large number of events is sampled and also the time duration of sampling is shorter than the duration associated with the spectral gap (order of 90 minutes). This requirement is generally satisfied by choosing a time length for statistical averaging that is somewhere between 30 and 60 minutes, with lower values associated with lower measurement heights and/or higher wind speeds; for a complete treatment, the reader is referred to Wyngaard (1973).

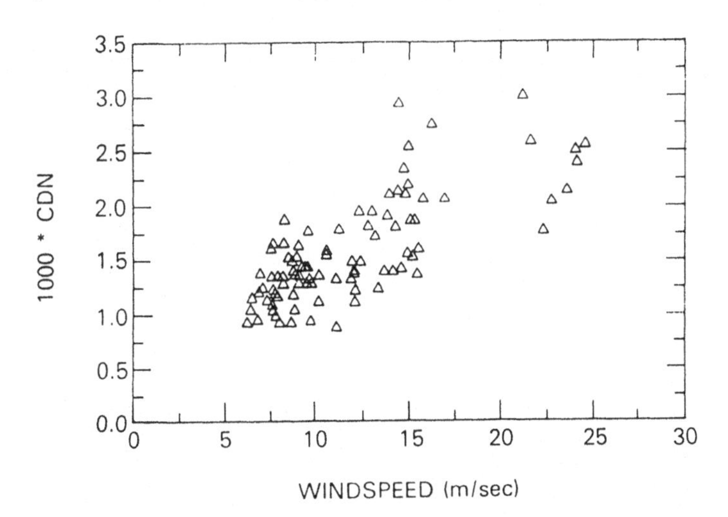

Figure 1: Distribution of the neutral drag coefficient at 10m elevation with windspeed. Data are from the 1985 FPN Experiment (Geernaert, et al., 1987).

To examine the periodicity of stress at a single point, measurements of the turbulent kinetic energy (TKE) dissipation rate are displayed as a time series in Figure 2. The dissipation rate, ϵ, is balanced by the production of TKE, which in turn scales with the stress according to:

$$\epsilon = u*^3 \phi_\epsilon / kz \tag{14}$$

where ϕ_ϵ is a stratification function (unity for neutral, and greater than one for diabatic conditions). The dissipation method has been widely used for estimating the over-ocean wind stress; see, e.g., Geernaert, et al., 1988b; Fairall and Larsen, 1984; and Guest and Davidson (1987). Note that periodicities of 3 to 5 minutes are readily evident. Note also in Figure 2 that periodicities (though less obvious) also appear in the stress signal as well as in the radar cross-section time series. The angle between the stress and wind are periodic at much higher frequencies. All four parameters plotted in Figure 2 represent 30 minutes of sampling, a length of time typical over which mean wind stress and drag coefficient calculations are performed; note especially the strong variability of all the signals.

To get an estimate of typical time scales associated with the boundary layer, we must first make use of boundary layer depth estimates based on surface stress scaling for neutral stratifications (e.g., Tennekes, 1973):

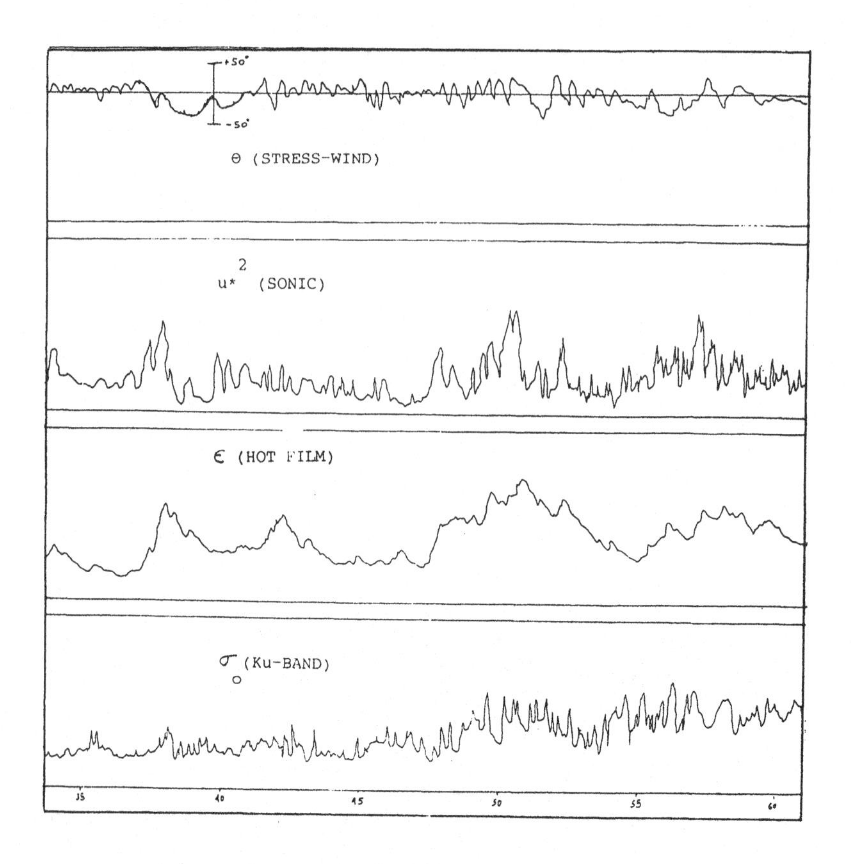

Figure 2: Typical variability of meteorological turbulence signals during a 30-minute time series. From top to bottom: angle between the stress and wind; directly measured wind stress with a sonic anemometer; turbulent kinetic energy dissipation rate; and Ku-band radar cross-section. Data were are from the TOWARD experiment; wind is 7 m/sec, and stratifications are slightly unstable, relative units and time is minutes past 1400 on March 17, 1985.

$$h = cu^*/f \tag{15}$$

where h is the PBL depth, f is the Coriolis parameter (order of 10^{-4} sec^{-1}), and the coefficient c is in the neighborhood of 0.3. Assuming that the largest eddy is circular and rotates with a period of $\pi h/U$, where U is the mean PBL wind speed, we arrive at a characteristic time scale, T, of

$$T = 0.3\ \pi f C_D^{1/2} \tag{16}$$

Since C_D is on the order of 10^{-3}, T is on the order of 5 minutes. For stable stratification, T will be somewhat smaller because C_D reduces dramatically with increasing stability. In consideration of the range of stratifications encountered over the ocean (particularly in the coastal zone where diurnal winds often prevail), one can expect realistic values of T to span from 3 to 6 minutes. Referring again to Figure 2, one may see strong periodicities in the neighborhood of 5 minutes.

Since lower frequency wind speed fluctuations on the 3-5 minute time scale have an implicit relationship with stress variations (through the bulk aerodynamic relationship), it follows that the resulting changes in the surface short wave slopes should be detectable with the same periodicity in a time series of radar cross-section at Bragg incidence angle. Referring again to Figure 2, cross-section measurements similarly exhibit low frequency periodicities that correspond to the 3-6 minute time scale associated with PBL circulations.

While wind stress magnitudes readily indicate periodicities based on boundary layer scales, no variability in the direction of the stress vector on these scales has yet been investigated. On longer scales, however, there has been one attempt, i.e., the study reported by Geernaert (1988), which compiled previously analyzed wind stress data in order to identify patterns or periodicities in the direction of the wind stress vector. While a clear stratification influence was found during conditions when both the wind speeds were high and the surface layer was well-mixed, data collected during a long period of very stable flow indicated lower frequency periodicities that were presumably driven by inertial oscillations. Referring to Figure 3, one can easily visualize a sinusoidal periodicity on the order of 3 or 4 hours. In order to explain such phenomenon, one needs to examine the Navier-Stokes equations in their simplified form, in this case neglecting the stress terms due to the very stable conditions, i.e.,

$$d(U-U_g)/dt = f(V-V_g) \tag{17}$$

$$d(V-V_g)/dt = -f(U-U_g) \tag{18}$$

where U_g and V_g are the Cartesian components of the geostrophic velocity. In (17) and (18), we assume that $dU_g/dt = dV_g/dt = 0$. Differentiating (17) and (18) and combining, one obtains a periodic solution of the form:

$$(U-U_g, V-V_g) = (U_0, V_0) \sin ft \tag{19}$$

where (U_0, V_0) represents mean states, and U and V become periodic with a time scale of $2\pi/f$, i.e., the inertial time scale. It must be pointed out that this time scale is with respect to the moving air mass. Taking into account advection, the time scale T observed at a fixed site on the surface must be determined by including advection, i.e.,

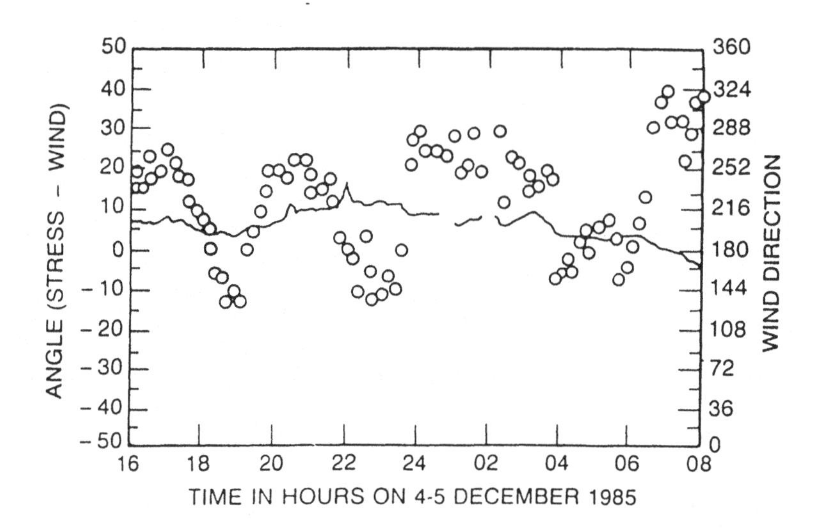

Figure 3: Time series of the angle between the stress and wind vectors measured with a sonic anemometer on a boom from the North Sea Platform in December 1985. Wind speeds are 6 m/sec with stable stratifications. Circles are ten-minute angle measurements while the line represents the wind direction.

$$2\pi/T = f + (K\ U) \tag{20}$$

where K represents the air mass wavenumber. The data in Figure 3 were determined to have an estimated intrinsic value of T near 4.6 hours which is close to the observed periodicity.

While both the boundary layer eddy induced periodicities in the stress magnitude and the stress vector variations based on inertial oscillations exhibit important features in mesoscale remote sensing interpretations, the problems of wave/wave interaction, wave/current interaction, stress variation as a function of long wave phase must be known. Little is known on this topic, but we intend to pose new insights in the next section.

4. SMALL SCALE PERIODICITIES

It is well-established that the energy and slope of short waves on the ocean surface have a statistical distribution along the phase of propagating long waves. Characteristically, short waves tend to congregate on the upwind face of the long waves with minima in the troughs (for long waves that travel slower than the wind). When analyzing the development and growth of these short waves, it has generally been assumed that the stress is more or less constant or at least varying insignificantly when compared to other important processes that affect wave dynamics. However, regardless of the application of stress modulation over long waves (see, e.g., the discussion by Longuett-Higgins, 1960), a better understanding of the stress modulation is needed to gain insight on how to better parameterize wave effects on the drag coefficient when the wave directional spectrum includes both a wind wave field and swell from many direction.

Three important points must be considered to tackle the wind stress wave state problem. First, roughness elements on the surface are periodic with important long waves. These long waves include the peak of the equilibrium range and large swell, where the common feature of these long waves is that they have considerable slope and/or large orbital velocities. The second point considers the spatial variation in the wind shear near the surface. The wind shear is most easily identified by assuming that the wind speed is more or less uniform (in the average) a few wave heights above the interface while the interface exhibits a periodicity induced by the long wave orbital velocities. In this case, the wind shear will become periodic with respect to the phase of dominant surface wavelengths, particularly if the wave age is less than unity. The third point is that the wave crest will be more exposed to the free stream wind speed than will the trough. Along these three lines of thought, the wind stress will depend on the distribution of roughness elements, their relative speeds with respect to the wind, and similarly their elevation above the surface. We will now treat each of these independently.

We can make a first order guess of the wind shear periodicity by first assuming that on the ocean surface the wave age is simply $c_o/u{*}=20$ (for illustration only). With this, the peak frequency, ω_o, of the equilibrium range can be approximated to be 1.4 g/U_{10}. Given that the wave amplitude, A_o, is twice the r.m.s. wave height, the orbital velocity projection in the wind direction, V_o, may be approximated as:

$$V_o = 2.5\ u{*} \cos\ (2\pi x/\lambda_o) \cos\ (\theta_w - \theta_o) \tag{21}$$

where θ_w and θ_o are respectively the direction of the wind and waves. Note that the coefficient "2.5" in (21) is appropriate only for open ocean waves, and this coefficient will be considerably smaller for wave tank wind and wave studies.

By first ignoring swell, and noting that a mean wind drift current has a value on the order of u* (Hicks, 1973), we can write a periodic equation for the surface velocity projection, U_{sfc}, into the direction of the wind. By including the projected tidal speed, U_t, we now have:

$$U_{sfc} = u{*} + U_t + 2.5\ u{*} \cos\ (2\pi X/\lambda_o) \tag{22}$$

where we have now assumed that the wind waves have the same direction as the wind

At this point, we can simulate the effect of surface velocity periodicities on the stress by rewriting (10) into the form:

$$u*^2 = A\ C_D\ U_{10}^2 \tag{23}$$

where

$$A = (1 - 2\ U_{sfc}/U_{10}) \tag{24}$$

and "A" represents a periodic coefficient in the neighborhood of unity that is based on long wave orbital variations. Combining (22) with (24), one may approximate A to be:

$$A = 0.94 - 2\ U_t/U_{10} + 5\ C_D^{1/2}\ \cos(2\pi x/\lambda_o) \tag{25}$$

Noting that C_D has a magnitude on the order of 10^{-3}, "A" will exhibit a periodic variation on the order of 15% to 20% according to phase of the long wave. Note again that the coefficient "5" in (25) is applicable only to open ocean waves of long fetch and duration.

While the above analysis of wind shear variation due to surface speed modulation exhibits a 15 to 20% variability in stress, we must also consider the additional variability of wave roughness according to phase of the long wave. To approach this problem, we must assume some relation between the roughness length z_o and wave state. Two approaches may be considered: (a) simply assume that the stress is due in large part to the mean slope of short waves (Phillips, 1980); or use a parameterization of the wind drag according to a model such as Kitaigorodskii's (1973) approach. In either case, we must have prior information on the distribution of wave slopes according to phase of long waves within the higher wave-number part of the wave spectrum. For a review of theoretical and experimental measurements of the modulation transfer function applicable to short waves in the presence of long waves, we refer to Plant (1988). For illustration only, we will assume that short waves in the gravity-capillary range exhibit a 25% to 50% modulation in slope within the developing to fully-developed wind sea (in the open ocean), where these high percentages are attached to wavenumbers in the capillary range. If the higher frequency gravity waves and capillary waves actually support the bulk of the stress, an argument posed by Geernaert (1983), then one may assume that the stress is modulated by nearly the same amount. One may alternatively apply Kitaigorodskii's relation, i.e.

$$z_o = .02\ [\int_0^\infty S(\omega)\ \exp(-2kc/u*)\ d\omega]^{1/2} \tag{26}$$

where $S(\omega)$ represents the spectral estimates of wave energy. In general, the primary contributions to the integral in (26) for waves approaching full development at moderate wind speeds are in the gravity and gravity-capillary range of the wave spectrum. Assuming that these "important" waves have slopes that vary in the mean by 50% over the lengths of the long waves, the variation in z_o and C_D can be easily calculated to be 22% and 3%, respectively. Clearly, there is a large disparity between the two predictions. It is noteworthy to point out that the greatest short wave slopes (which suggest greater drag) will be found near the crests of the long waves. Recall that the crests will exhibit higher surface speeds which correspond to lower wind shear and lower drag, thereby cancelling to some degree the effect due to short wave modulation.

Experimental evidence for wind stress modulation over the open ocean is sparse, and when collected, the physical meaning of the data is difficult to resolve. Direct wind stress measurements collected during MARSEN were used to both determine the magnitude of the drag coefficient (Geernaert, et al., 1986) and also to evaluate wave effects on the momentum flux cospectrum (Geernaert, 1983). The stress cospectra calculated by Geernaert (1983) were normalized so as to give a "flat" appearance in the mean (see Figure 4), and significant deviations from this mean would in part indicate wave effects. Referring to Figure 4, many deviations occurred during the time series, but repeatable peaks appeared at frequencies that corresponded to the dominant surface wave frequencies. Two peaks predominated: the spectral peak corresponding to the peak of the wind wave spectrum, and a spectral peak associated with swell propagating in from a different quadrant of the storm. The matching of stress cospectral peaks with surface wave peaks suggests a strong coupling, but we are currently unable to properly explain the dynamics of the interaction. We can only speculate that a considerable modulation of the stress magnitude will exist in the presence of storm waves; a modulation of the direction of the stress vector about a mean direction may also exist, especially if there exists a large angle between the mean wind and the direction of long wave propagation. The SAXON experiment, to be conducted in September 1988, is designed to address the issues of stress vector modulation by long waves and swell, especially for cases when the swell, wind waves, and wind have significantly different directions.

5. SUMMARY

This review is intended to place into perspective the variability of the stress vector on short time scales (tens of meters) and longer boundary layer and inertial scales (one to tens of kilometers). While much attention has been focussed on understanding the statistical behavior of the stress on the longer time and space scales (for applications to numerical prediction models of the atmosphere and ocean), little attention has been paid to shorter term variations (for application to wind-wave dynamics and surface layer turbulence). With this article, we hope that new ideas and research will be spawned.

6. ACKNOWLEDGEMENTS

This research was supported by the Basic Research Program of the U.S. Naval Research Laboratory. Since the data presented herein represent

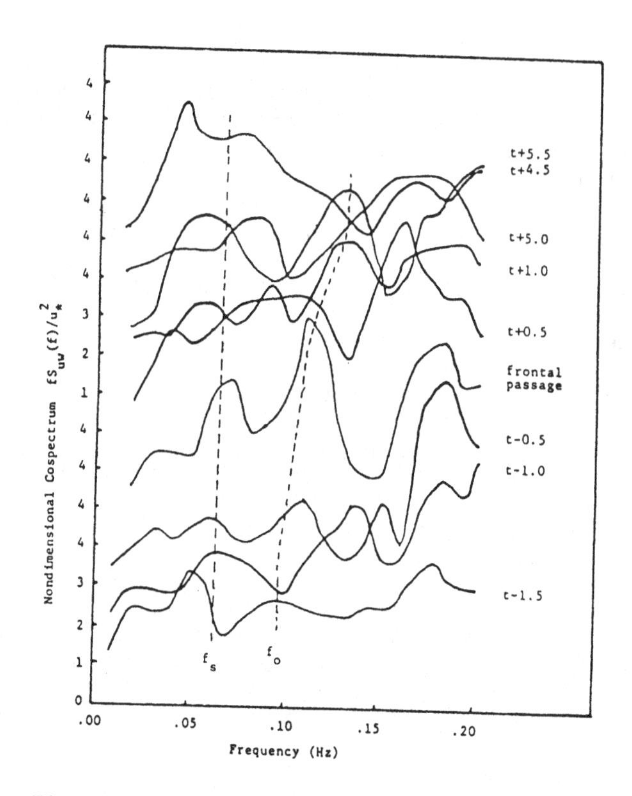

Figure 4: Time series of the momentum flux cospectrum (lower frequencies only) during the passage of a cold front over the North Sea. Data were collected from the Pisa needle with a Gill anemometer during the MARSEN experiment. The local wind sea peak frequency and swell frequency are denoted by f_o and f_s (in Hz), respectively.

the TOWARD, MARSEN, and 1985 North Sea Platform Winter Experiments, special acknowledgements are also due to Ken Davidson (Naval Postgraduate School, Monterey, CA), Soren Larsen and Torben Mikkelsen (Riso National Laboratory, Denmark), Kristina Katsaros (University of Washington, Seattle, WA), and Bill Keller (Naval Research Laboratory, Washington, DC).

7. REFERENCES

Barger, W. R., W. D. Garrett, E. L. Mollo-Christensen, adn K. Ruggles, Effects of an artificial slick upon the atmosphere and ocean, J. Appl. Met., 9, 396-400, 1970.

Busch, N. E., On the mechanics of atmospheric turbulence, in Workshop on Micrometeorology, ed. D. Haugen, Amer. Meteorolog. Soc., Boston, 1973.

Businger, J. A., J. C. Wyngaard, and Y. Izumi, Flux profile relationships in the atmospheric surface layer, J. Atmos. Sci., **28**, 181-189, 1971.

Businger, J. A., Turbulent transfer in the atmospheric surface layer, in Workshop on Micrometeorology, D. Haugen, ed., Amer. Meteorol. Soc., Boston, 1973.

Charnock, H., Wind stress on a water surface, Q. J. Roy. Met. Soc., **81**, 639-640, 1955.

Donelan, M. A., The dependence of the aerodynamic drag coefficient on wave parameters, in First International Conference on Meteorology and Oceanography of the Coastal Zone, pp. 381-387, Amer. Met. Soc., Boston, 1982.

Donelan, M. A., The effect of swell on the growth of wind waves, NWRI Contrib. No. 86-117, NWRI, Canada Center for Inland Waters, Burlington, Ontario, 1986.

Fairall, C. W., and S. E. Larsen, Inertial-dissipation methods and turbulent fluxes at the air-ocean interface, Bound. Layer Met., 34, 287-301, 1984.

Geernaert, G. L., Variation of the drag coefficient and its dependence on sea state, Ph.D. Dissertation, Univ. of Washington, Seattle, 209 pp, 1983.

Geernaert, G. L., K. B. Katsaros, and K. Richter, Variation of the drag coefficient and its dependence on sea state, J. Geoph. Res., **91**, 7667-7679, 1986.

Geernaert, G. L., and K. B. Katsaros, Incorporation of stratification effects on the oceanic roughness length in the derivation of the neutral drag coefficient, J. Phys. Oceanogr., **16**, 1580-1584, 1986.

Geernaert, G. L., S. E. Larsen, and F. Hansen, Measurements of the wind stress, heat flux, and turbulence intensity during storm conditions over the North Sea, J. Geoph. Res., **92**, 13127-13139, 1987.

Geernaert, G. L., K. L. Davidson, S. E. Larsen, and K. Mikkelsen, Measurements of the wind stress during the Tower Ocean Wave and Radar Dependence Experiment, J. Geoph. Res., in press, 1988.

Geernaert, G. L., Observations of the direction of the wind stress vector with respect to the direction of the mean surface layer flow over the North Sea, J. Geoph. Res., in press, 1988.

Guest, P. S., and K. L. Davidson, The effect of observed ice conditions on the drag coefficient in the summer East Greenland Sea Marginal Ice Zone, J. Geoph. Res., **92**, 6943-6954, 1987.

Hasselmann, K., R. P. Barnett, E. Bouws, H. Carlsen, D. E. Cartwright, A. Meerburg, P. Muller, D. J. Olbers, K. Richter, W. Sell, and H. Walden, Measurements of wind wave growth and swell decay during the Joint North Sea Wave Project (JONSWAP), Erganzugsh. Dtsch. Hydrog. Z., 12A(i), 1975.

Hicks, B. B., Some evaluations of drag and bulk transfer coefficients over water bodies of different sizes, Bound. Layer Met., 3, 201-213, 1973.

Hsu, S. A., A dynamic roughness equation and its application to wind stress determination at the air-sea interface, J. Phys. Oceanogr., 4, 116-120, 1974.

Huang, N. E., L. F. Bliven, S. R. Long, and P. S. DeLeonibus, A study of the relationship among wind speed, sea state, and the drag coefficient for a developing wave field, J. Geoph. Res., **91**, 7733-7742, 1986.

Kitaigorodskii, S. A., The Physics of Air-Sea Interaction, translated from Russian by A. Baruch, Israel Program for Scientific Translations, Jerusalem, 1973.

Kitaigorodskii, S. A., and V. A. Razumov, Interactions of currents with waves, Izv. Atmos. Ocean Physics, 14, 639-642, 1978.

Longuett-Higgins, M. S. and R. W. Stewart, Changes in the form of short gravity waves on long waves and tidal currents, J. Fluid Mech., **52**, 565-583, 1960.

Melville, W. K., Wind stress and roughness length over breaking waves, J. Phys. Oceanogr., **7**, 702-710, 1977.

Phillips, O. M., The Dynamics of the Upper Ocean, Cambridge Univ. Press, 336 pp, 1980.

Plant, W. J., A relationship between wind stress and wave slope, J. Geoph. Res., **87**, 1961-1967, 1982.

Plant, W. J., The modulation transfer function: concept and applications, this volume, 1988.

Smith, S. D., Wind stress and heat flux over the ocean in gale force winds, J. Phys. Oceanogr., **10**, 709-726, 1980.

Tennekes, H., Similarity laws and scale relations in planetary boundary layers, in Workshop on Micrometeorology, ed. D. Haugen, Amer. Met. Soc., Boston, 177-214, 1973.

Wyngaard, J. C., On surface layer turbulence, in Workshop on Micrometeorology, Amer. Met. Soc., Boston, 101-149, 1973.

Zhang, S., Evaluation of the von Karman constant, Ph.D. dissertation, Univ. of Washington, Seattle, 1988.

ENERGY BALANCE IN SMALL-SCALE WAVES — AN EXPERIMENTAL APPROACH USING OPTICAL SLOPE MEASURING TECHNIQUE AND IMAGE PROCESSING

BERND JÄHNE
Scripps Institution of Oceanography
University of California
La Jolla, CA 92093-0212
USA

ABSTRACT. The energy balance in small-scale waves is investigated with optical wave-slope measuring technique. Time series from one point at the water surface are obtained with a laser slope gauge and image sequences of the wave slope with an imaging slope gauge. Frequency spectra, wavenumber spectra, and wavenumber-frequency spectra are calculated from data obtained in different wind/wave facilities. They show a strong nonlinearity of the waves already for wind speeds as low as 1.9 m/s. When the dominant gravity wave frequency is smaller than 4 Hz, an equilibrium range establishes for waves with frequencies larger than 10 Hz. In this range, the slope frequency spectrum is proportional to $u_\star^{5/2}$, but still increases with fetch, while small gravity waves with frequencies larger than the dominant wave show only very little dependence on fetch and $u_\star$. These experimental facts indicate that the energy balance in small-scale waves is more complex than assumed by previous theories.

1 Introduction

Short gravity and capillary waves play an important role in small scale air-sea interaction. They are critical parameters for the transfer of heat, mechanical energy, momentum, and gases across the air-ocean interface [*Coantic*, 1980; *Jähne et al.*, 1987]. In addition, they receive increasing interest because of their significance for radar backscattering.

In this paper a detailed experimental study of the spectral characteristics of small-scale wind waves based on slope measurements is presented aiming at a better understanding of the energy balance in small-scale waves. The study includes frequency, wavenumber and wavenumber-frequency spectra.

First, the energy balance in small-scale waves and the advantage of slope measurements are briefly discussed. Section 3 describes the optical slope measuring technique. Finally, experimental results from several wind/wave tunnel studies are presented and discussed in the last section.

G. J. Komen and W. A. Oost (eds.), Radar Scattering from Modulated Wind Waves, 105–120.

2 Energy Balance in Small-Scale Waves

2.1 Saturation Range

The simplest concepts of an upper-limit asymptote of the spectrum, the *saturation range*, goes back to the early work of [*Phillips*, 1958]. He postulated that the dissipation of waves (e.g. by wave breaking) imposes an upper limit to the spectral density. If this limit is independent of the energy input by the wind and only depends on the restoring gravity and capillary forces, the following spectral densities are obtained for the height spectra

$$\Psi_s(\boldsymbol{k}) \;\propto\; f(\Theta)\,k^{-4} \qquad \begin{array}{lcll} \Phi_s(\omega) & \propto & g^2\,\omega^{-5} & \text{gravity waves} \\ \Phi_s(\omega) & \propto & \gamma^{2/3}\,\omega^{-7/3} & \text{capillary waves,} \end{array} \tag{1}$$

where γ is the surface tension σ over the density ρ and ω is the circular frequency. The angular dispersion is expressed in the function $f(\Theta)$, the index s marks the saturation spectrum. Clearly, the concept of a saturation range is an oversimplification. Wave dissipation will not start if a critical amplitude is exceeded but rather increase gradually with amplitude. Nevertheless, the saturation range is a useful concept which will be used here to express the spectral densities in a dimensionless function, the *degree of saturation*, B, as proposed by *Phillips* [1985]

$$B(\boldsymbol{k}) = \frac{\Psi(\boldsymbol{k})}{\Psi_s(\boldsymbol{k})} \quad B(\omega) = \frac{\Phi(\omega)}{\Phi_s(\omega)}. \tag{2}$$

2.2 Equilibrium range

A more detailed analysis shows that the energy balance of the waves includes the following terms [*Phillips*, 1985]:

- Energy input by the turbulent wind field;
- Transfer of energy between waves of different wavenumbers by nonlinear wave-wave interaction;
- Dissipation of energy by wave breaking, viscous and turbulent diffusion.

Using the degree of saturation B, *Phillips* [1985] derives the following terms for the energy fluxes expressed as fluxes of the action spectral density for gravity waves:

$$\begin{array}{rcl} \text{wind input} & = & m\,f(\Theta)\,gk^{-4}\left(\frac{u_*}{c}\right)^2 B(\boldsymbol{k}) \\ \text{spectral flux divergence} & \propto & gk^{-4}B^3(\boldsymbol{k}) \\ \text{dissipation} & = & gk^{-4}\,f(B(\boldsymbol{k})), \end{array} \tag{3}$$

where u_* is the friction velocity in air and c the phase velocity of the waves. While the two first processes are rather well known, the dissipation term still has to be investigated. Even the dominant mechanism for dissipation is not yet known. Only very recently, turbulent diffusion has been considered as a dissipation source [*Rosenthal*, 1988].

Another big problem is related to the fact that quite different assumptions about the energy balance lead to the same spectral densities. In other words, inferring the energy balance from spectral densities is an ill-defined inverse problem. Two examples:

On the one hand, *Phillips* [1985] assumes that all three fluxes in (3) are of equal importance. A local balance yields

$$\begin{array}{rcl} \Psi_e(\boldsymbol{k}) & = & \beta\,f(\Theta)\,\left(\frac{u_*}{c}\right)k^{-4} = \beta\,f(\Theta)\,u_*\,g^{-1/2}\,k^{-7/2} \\ \Phi_e(\omega) & = & \alpha u_* g\omega^{-4}. \end{array} \tag{4}$$

On the other hand, *Kitaigorodskii* [1983] proposed the existence of a Kolmogoroff-type energy cascade in which the wind input is assumed to occur primarily at the energy containing large scales and dissipation at small scales. Then, over a range of wavenumbers, all fluxes in (3) are negligible and the spectral energy flux, ϵ_0, is constant in this range. If ϵ_0 is proportional to u_*^3, these considerations yield exactly the same spectral densities.

Similar considerations about the energy balance for gravity-capillary and capillary waves are lacking. As a first estimate, we may take the same fluxes of the spectral densities as for gravity waves (3). Only the spectral flux divergence due to nonlinear wave-wave interaction has to be chanced. Since interactions between gravity-capillary waves take place in triplets instead of quartets, we assume this term to be rather quadratic than cubic in the degree of saturation. Using the linear dispersion relation for capillary waves $\omega^2 = \gamma k^3$ yields

$$\begin{aligned} \Psi_e(\boldsymbol{k}) &= \beta' f(\Theta) \left(\tfrac{u_*}{c}\right)^2 k^{-4} = \beta' f(\Theta)\, u_*^2 \gamma^{-1} k^{-5} \\ \Phi_e(\omega) &= \alpha' u_*^2 \omega^{-3}. \end{aligned} \tag{5}$$

In these equations viscous damping is neglected. Capillary waves are more sensitive to friction velocity than gravity waves ($\propto u_*^2$ in contrast to $\propto u_*$).

In order to gain more insight into the energy balance, it is necessary to obtain more experimental data under a wide range of conditions. Studies under controlled conditions, i.e. in wind-wave facilities, seem to be most suitable. This paper concentrates on wave slope measurements. Though slope measuring optical techniques have been used by several authors [*Cox*, 1958; *Long and Huang*, 1976; *Reece*, 1978; *Lubard et al.*, 1980; *Tang and Shemdin*, 1983], the data basis is still poor. It is the intention of this paper to improve knowledge about the energy balance of small-scale waves by (a) comparing results from different wind-wave facilities and (b) measure both frequency and wavenumber spectra.

3 Slope Spectra

Wave slope is a vector with two components, the along-wind component s_1 and the cross-wind component s_2, since it is the space derivative of the water surface deflection a

$$\boldsymbol{s}(\boldsymbol{x},t) = \nabla a(\boldsymbol{x},t). \tag{6}$$

The relation between the slope and amplitude power spectra can easily be deduced from the Fourier transform of the water surface deflection

$$a(\boldsymbol{x},t) = \int_{\mathbf{k}} \int_{\omega} \hat{a}(\boldsymbol{k},\omega) \mathrm{e}^{\mathrm{i}\mathbf{kx}-\omega t} \mathrm{d}\boldsymbol{k} \mathrm{d}\omega, \tag{7}$$

where $\hat{a}(\boldsymbol{k},\omega)$ is the complex valued amplitude function.

$$\begin{aligned} S_1 &= k_1^2 |\hat{a}(\boldsymbol{k},\omega)|^2 = k^2 \cos^2\Theta\, A(\boldsymbol{k},\omega) \\ S_2 &= k_2^2 |\hat{a}(\boldsymbol{k},\omega)|^2 = k^2 \sin^2\Theta\, A(\boldsymbol{k},\omega) \\ S = S_1 + S_2 &= k^2 |\hat{a}(\boldsymbol{k},\omega)|^2 = k^2 A(\boldsymbol{k},\omega). \end{aligned} \tag{8}$$

$A(\boldsymbol{k},\omega)$ is the directional wavenumber-frequency spectrum of the wave amplitude and Θ the angle between the wind direction and the propagation direction of the wave.

3.1 Saturation Slope Spectra

The spectral densities of the wave slope spectra have to be multiplied by k^2 as compared with the corresponding height spectra. Using the corresponding (linear theory) dispersion relations yields the following simple saturation ranges for the slope spectra from (1) which can also be derived directly by dimensional analysis

$$\begin{array}{rclrcl} S_1(\boldsymbol{k}) & \propto & \cos^2\Theta\, f(\Theta)\, k^{-2} & S_1(\omega) & \propto & \langle\cos^2\Theta\rangle\, \omega^{-1} \\ S_2(\boldsymbol{k}) & \propto & \sin^2\Theta\, f(\Theta)\, k^{-2} & S_2(\omega) & \propto & \langle\sin^2\Theta\rangle\, \omega^{-1} \\ S(\boldsymbol{k}) & \propto & f(\Theta)\, k^{-2} & S(\omega) & \propto & \omega^{-1}, \end{array} \tag{9}$$

where $\langle\cdots\rangle$ is the mean value weighted with the angular dispersion function $f(\Theta)$. Surprisingly, there is no longer a difference between capillary and gravity waves in the frequency spectra. This fact emphasizes the slope as a basic parameter for the waves. The ω^{-1} and k^{-2} power laws mean that all waves range up to a constant slope in (more natural) logarithmic intervals ($\mathrm{d}\omega\,\omega = \mathrm{d}\ln\omega$ and $\mathrm{d}\boldsymbol{k}\,k^2 = \mathrm{d}\ln\boldsymbol{k}$). Since the limiting slope for capillary and gravity waves is similar, also approximately the same constants are expected.

From the experimental point of view, slope measurements offer a significant advantage over height measurements. Any height gauge must cover a signal range of several decades in order to measure small ripples riding on top of larger gravity waves, but both types of waves have similar slopes. Consequently, any slope measuring instrument needs only a low dynamics to cover the whole range of waves from large gravity waves to the smallest ripples.

3.2 Equilibrium Slope Spectra

The equations for the equilibrium spectrum for the total slope spectra follow directly from (4) and (5)

$$\underbrace{\begin{array}{rcl} S_e(\boldsymbol{k}) & = & \beta\, f(\Theta)\, u_\star\, g^{-1/2}\, k^{-3/2} \\ S_e(\omega) & = & \alpha u_\star g^{-1} \end{array}}_{\text{gravity waves}} \quad \underbrace{\begin{array}{rcl} S_e(\boldsymbol{k}) & = & \beta'\, f(\Theta)\, u_\star^2\, \gamma^{-1}\, k^{-3} \\ S_e(\omega) & = & \alpha' u_\star^2 \gamma^{-2/3} \omega^{-5/3}. \end{array}}_{\text{capillary waves}} \tag{10}$$

As for the height spectrum, the spectral densities are proportional to $u_\star$ and $u_\star^2$, respectively. While the degree of saturation is increasing for gravity waves with $k^{1/2}$ and ω, respectively, it is decreasing for capillary waves again with k^{-1} and $\omega^{-2/3}$, respectively.

4 Optical Wave Slope Measuring Technique

4.1 Principle: Light refraction

The optical instruments used to measure the wave slope are based on refraction of light at the water surface. A light beam piercing the water surface perpendicularly from above is refracted by a sloping water surface (Fig. 1a). The relation between the slope of the water surface $s = \tan\alpha$ and the angle of the refracted beam in the water, γ', is given by the refraction law to

$$\tan\alpha = s = \frac{n_w \tan\gamma'}{n_w - \sqrt{1+\tan^2\gamma'}} \approx 4\tan\gamma' \left(1 + \frac{3}{2}\tan^2\gamma'\right). \tag{11}$$

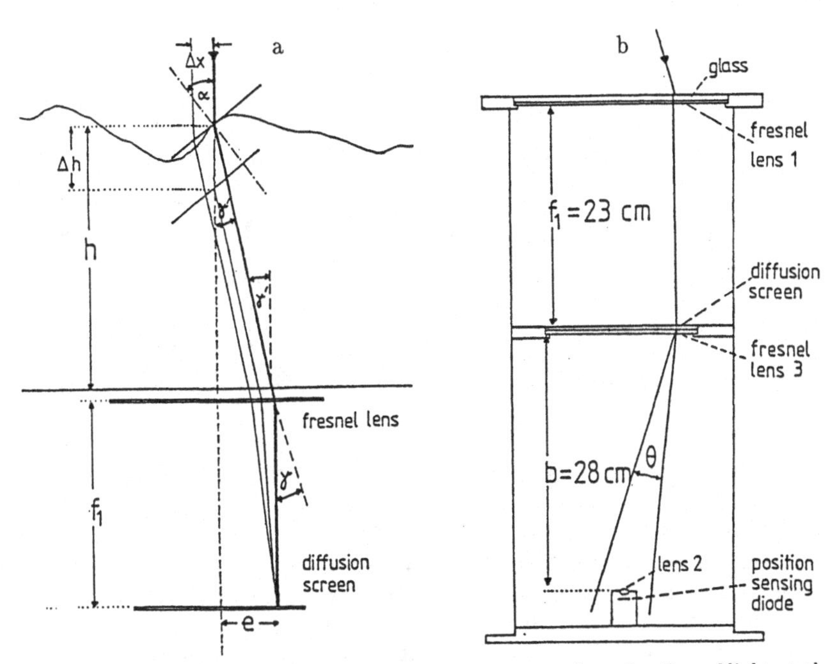

Figure 1: a) Basic optical geometry for wave slope measurements by refraction of light at the water surface; b) cross-section of the submerged optical receiver of the laser slope gauge.

(In the approximation for small $\tan\gamma'$ the refractive index of water, n_w, was taken as 4/3.) For slopes up to 1 ($\alpha = 45°$), the nonlinear terms in (11) are below 10%. In principle, slopes up to infinity can be measured. The maximum γ' is 41.4°.

4.2 Laser Slope Gauge

The light beam of a 5mW He-Ne laser vertically pierces the water surface. The beam which is tilted by refraction is collected by a submerged optical receiver (Fig. 1a). It consists of a Fresnel lens and a diffusor one focal length f distant from the lens. Therefore the displacement of the light spot on the diffusor is independent both of the wave height and the measuring position (Fig. 1a). It is only function of the wave slope. Because of the second refraction at the water-glass-air interface, the relation between the wave slope and the displacement e on the diffusor, expressed by $\tan\gamma = e/f$, is given by a slightly different expression [*Lange et al.*, 1982]

$$\tan\alpha = s = \frac{\tan\gamma}{\sqrt{n_w^2 + (n_w^2 - 1)\tan^2\gamma} - \sqrt{1 + \tan^2\gamma}} \approx 3\tan\gamma\left(1 + \frac{5}{8}\tan^2\gamma\right). \qquad (12)$$

The nonlinearity of the slope-displacement relation is even lower ($< 7\%$ for $s < 1$) than for the simple case of only one refraction at the water surface (11).

A second lens images the spot on the diffusor onto a dual-axis position sensing photodiode (Fig. 1b). The amplifier circuit enables the two position signals to be divided by the intensity in order to avoid systematic errors due to intensity variations.

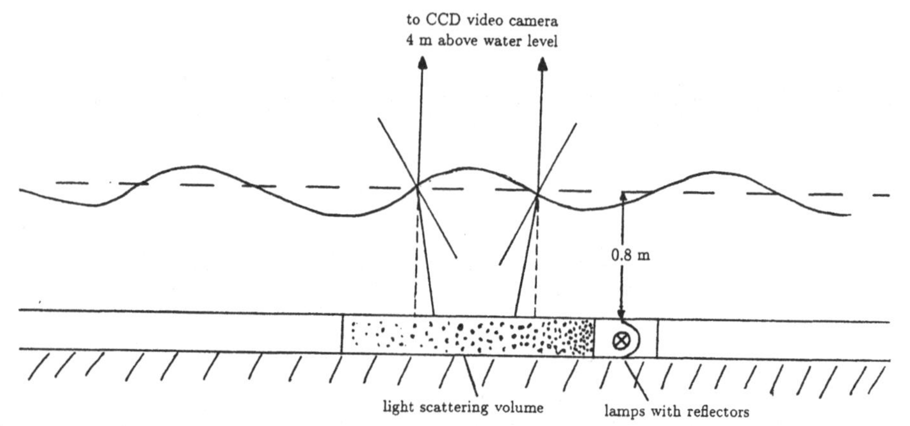

Figure 2: Outline of the imaging slope gauge (ISG) in the large wind/wave facility of Delft Hydraulics.

The systems allows the measurements of waves up to the highest frequency capillary waves without significant signal damping. The maximum measurable slope is 1.2. The instrument has been calibrated in situ. The small nonlinearities in the signal were corrected, but this correction showed no significant change in the spectral densities.

4.3 Imaging Slope Gauge

Wave image sequences can be taken with the imaging slope gauge. Compared to the LSG, the places of the light source and optical receiver are interchanged (Fig. 2). A CCD camera observes the water surface from the ceiling of the wind/wave facility. The submerged illumination system produces a horizontally changing intensity either in along- or cross-wind direction. This intensity change is compensated by an absorption wedge in front of the CCD-camera lens. Therefore an even water surface is imaged at approximately constant intensity all over the image sector. When the surface is sloped by waves, the light beam received by the CCD-camera comes from a place of higher or lower light intensity than for zero slope. In this way, the intensity is changing with the wave slope. Depending on the direction of the intensity change either the along-wind or cross-wind component of the wave slope is made visible. The gray value in the image is proportional to the corresponding wave slope component.

5 Results

In the last eight years, a number of wave slope measurements have been carried out in quite different wind-wave tunnels as listed in Table 1. These facilities cover a wide range of experimental conditions with fetches up to 90 m in linear tunnels. The two Heidelberg circular facilities represent an alternative approach for laboratory simulation offering the advantage of an unlimited fetch and an homogeneous wave field.

In all facilities slope frequency spectra have been measured using the LSG. In addition, wave image sequences have been recorded in the large wind-wave flume of Delft Hydraulics

Location	Water channel size			Maximum wind speeds [m/s]	Measurements
	Fetch [m]	Width [m]	Depth [m]		
Small circular facility University of Heidelberg	1.5[1]	0.1	0.05–0.07	6.9	LSG
Large circular facility University of Heidelberg	12[1]	0.3	0.05–0.35	12.1	LSG
Wind-wave flume University of Hamburg	25	1.0	0.5	16	LSG
Large wind-wave flume IMST, Marseille	40	2.6	0.7–1.0	13.8	LSG
Large wind-wave flume Delft Hydraulics	100	8.0	0.8	15	LSG, ISG

[1]circumference of annular water channel

Table 1: Basic features of wind/wave facilities used for wave slope measurements

within the frame of the VIERS-1 experiment in February 1988 [*Halsema et al.*, 1988].

5.1 Time series

Two examples of time series obtained with the LSG during the VIERS experiment are shown in Fig. 3. Already at the low wind speed of only 2.2 m/s the dominant wave of about 2.5 Hz are remarkably nonlinear. The slope decreases much faster than it increases indicating a larger curvature of the wave crests than of the wave troughs. In addition, the dominant waves shows considerable random fluctuations in shape, amplitude and frequency. Capillary wave trains with frequencies between 30 and 100 Hz occur in short bursts ($\approx$ 0.1 s) randomly spaced in time intervals of several seconds.

The 2 s time series at 8 m/s wind speed (Fig. 3b), clearly shows the modulation of the small scale waves in frequency and amplitude by the 1.1 Hz dominant gravity wave. The high frequency and steep capillary waves occur at the windward side of the gravity wave.

5.2 Slope Frequency Spectra

Slope frequency spectra have been measured in all wind-wave facilities listed in Table 1. The sampling frequency was 1 kHz. Time series with 4096 and 8192 (in the circular facilities with 32,768) values have been processed. The spectra shown are averages over typically 30 sample records with additional frequency averaging. All spectra are plotted in double logarithmic graphs. The spectral densities are multiplied by the frequency in order to obtain straight lines for the saturation range (9).

An overview over the data shows that the spectral densities do not follow a simple law. There are rather three different regimes:

Initial wave growth. In this range, the initially generated ripples of about 10 Hz peak frequency are amplified (Fig. 4a).

Bound capillary waves. The dominant wave is shifted towards lower frequencies and a secondary bump occurs in the capillary wave region between 50 and 100 Hz. As the dominant wave frequency is getting lower, the bump in the capillary region is shifted towards higher frequencies (Fig. 4b, c). This relation clearly indicates that the bump is caused by additional capillary waves directly generated by instabilities of the dominant gravity waves.

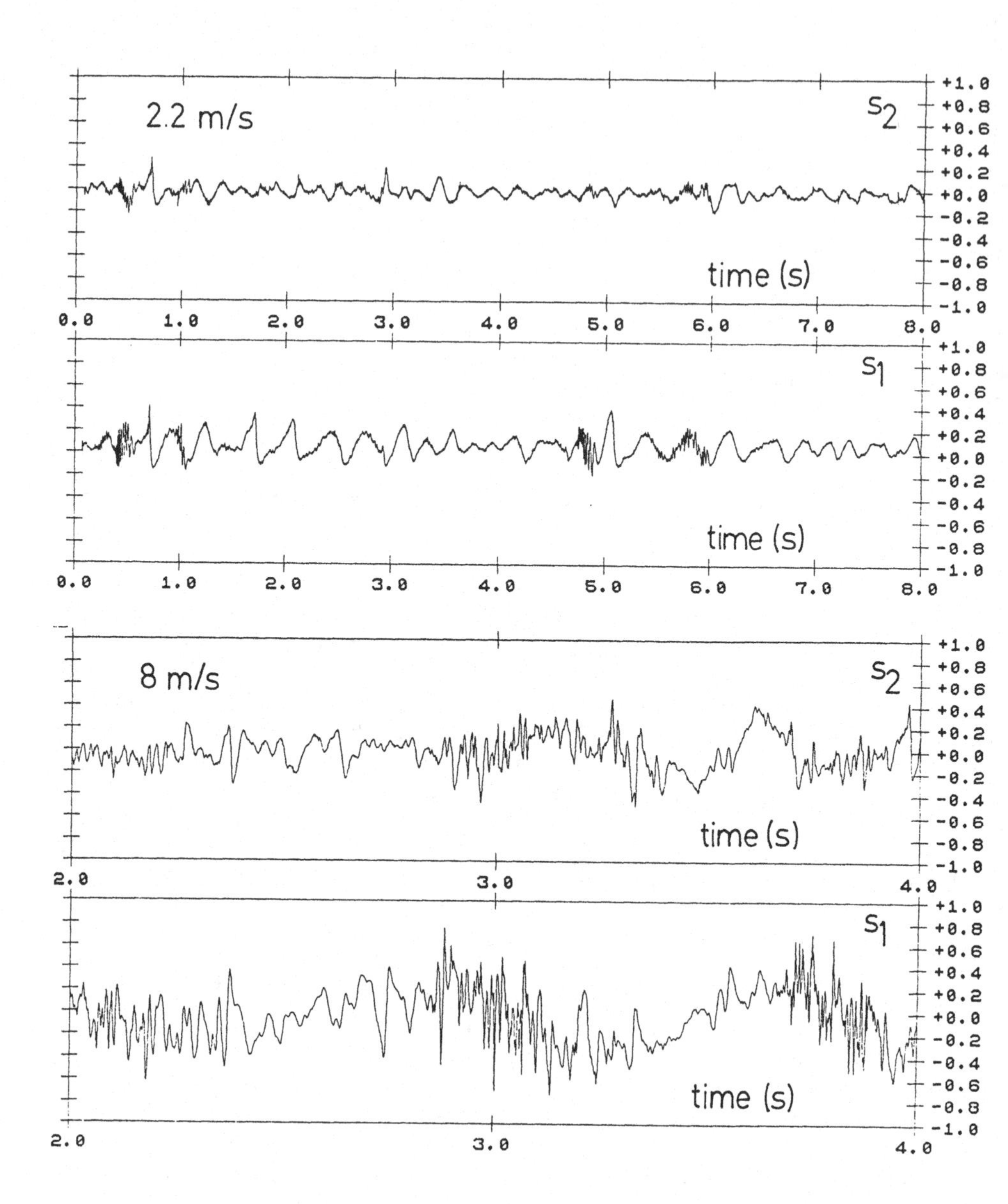

Figure 3: Time series of the along-wind slope s_1 and the cross-wind slope s_2 as obtained in the large wind-wave facility of Delft Hydraulics at 90 m fetch. Wind speeds as indicated; positive slopes are directed against the wind direction.

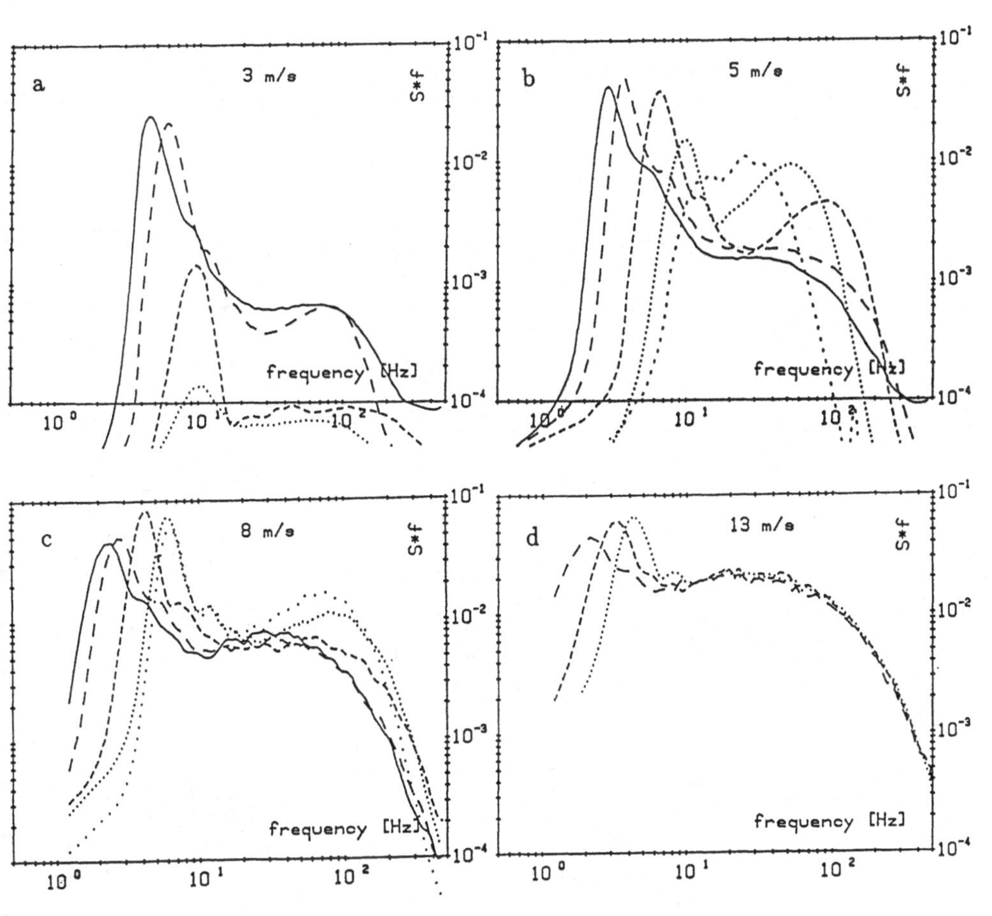

Figure 4: Fetch dependence of the total slope frequency spectrum as measured in the large wind-wave facility of the IMST, Marseille; wind speeds as indicated; water drift velocity 6 cm/s; key to fetch [m]: long-dotted: 2.6; short-dotted: 4.6; short-dashed: 9.4; long-dashed: 21; solid: 30.9.

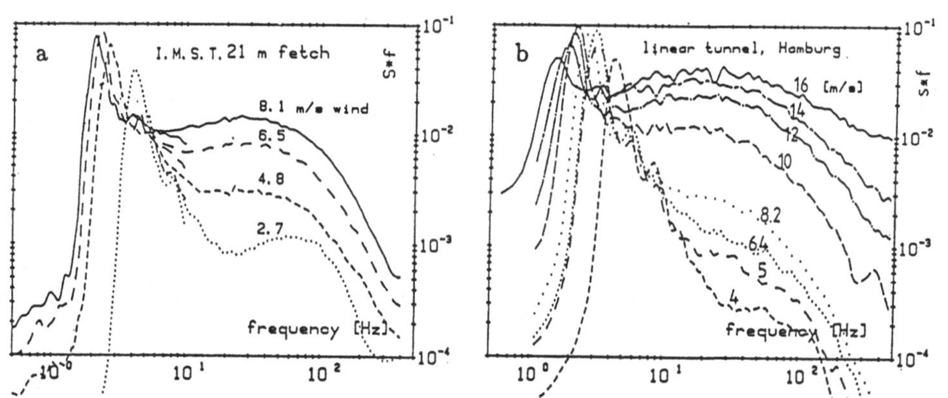

Figure 5: Wind speed dependence of the total slope spectrum: a) IMST facility, b) Hamburg wind tunnel; wind speeds in m/s as indicated.

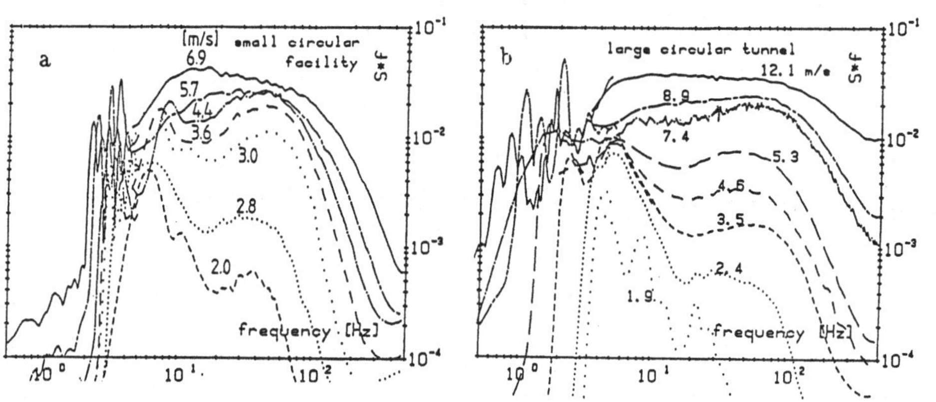

Figure 6: Wind speed dependence of the total slope spectrum: a) small circular facility, b) large circular facility of Heidelberg University; wind speeds in m/s as indicated.

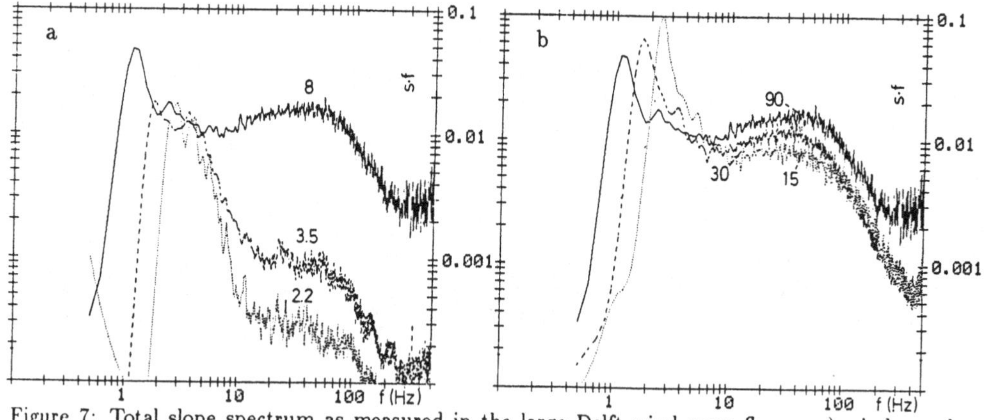

Figure 7: Total slope spectrum as measured in the large Delft wind wave flume: a) wind speed dependence at 90 m fetch; wind speeds in m/s as indicated; b) fetch dependence at 8 m/s wind, fetch in m as indicated.

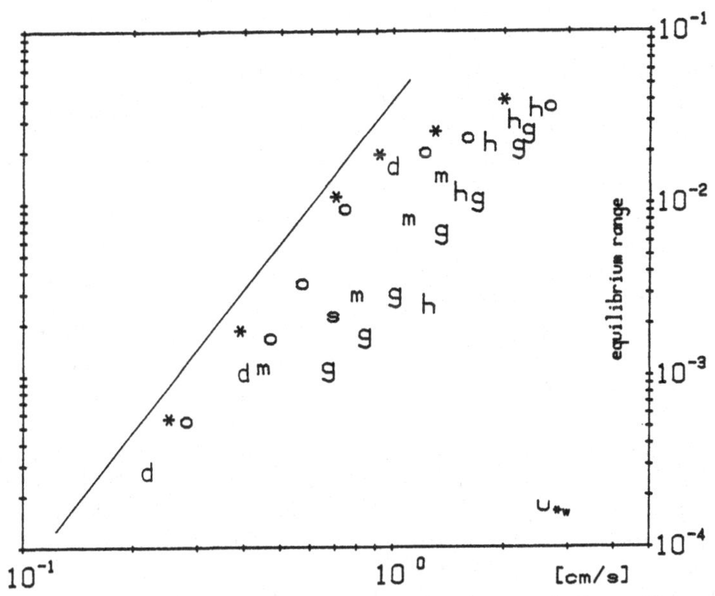

Figure 8: Spectral density of the total wave slope for small scale waves in the capillary-gravity range in different wind-wave facilities as a function of friction velocity in water $u_{\star w}$; stars: large circular facility; open circles: small circular facility; g: large IMST tunnel with 6 cm/s water drift velocity and a slight surface contamination; m: same tunnel, but without a drift current and with a clean water surface; h: Hamburg tunnel with a water surface contaminated by a surface film; s: same tunnel with a clean water surface; d: Delft wind/wave flume at 90 m fetch.

These waves have about the same phase speed as the dominant wave and therefore are bound at the lee-side of the gravity wave. This effect leads to at a strong overshoot in the spectral densities of capillary waves at low fetches up to one order of magnitude compared to the spectral densities at higher fetches (Fig. 4b, c).

Equilibrium range. When the frequency of the dominant wave is lower than about 4 Hz, the bump disappears and an equilibrium range establishes where the spectral densities of waves with frequencies larger than the dominant wave are independent of the fetch (Fig. 4b, c). However, the measurements in the much larger Delft facility show that a fetch independent equilibrium range seems to be limited to a rather small range. From 15 to 90 m fetch, the spectral densities nearly double at 8 m/s wind speed (Fig. 7b). The spectral densities $S(f)f$ are proportional to f^n, where $n = 0$ at high and low wind speeds and 0.25 at medium wind speeds up to frequencies of 50–100 Hz. Then a strong cut-off starts. These observations are in disagreement with the expected spectral shapes due to the energy balance considerations ($S(f)f \propto f^1$ for gravity and $\propto f^{-2/3}$ for capillary waves (10)).

In the range of small gravity waves (2–7 Hz), the spectral densities are remarkably independent of both fetch and wind speed when the dominant wave frequency is below that range (Fig. 7a and Fig. 5a), while a proportionality with $u_\star$ was expected (10).

However, as soon as capillary forces become important (beyond 10 Hz), the spectral densities rapidly start to be strongly dependent on the wind speed. Figure 8 shows the maximum spectral density in the capillary wave region as a function of the friction velocity in water, u_{*w}. In all facilities, the spectral densities initially increase with $u_*^{5/2}$. This is close to the postulated u_*^2 dependence for capillary waves (10). The steep increase gradually weakens. Asymptotically, a saturation level may be reached.

Surprisingly, the level of the equilibrium range is not the same in all facilities. The data show that it increases with fetch. The spectral densities are lowest in the Marseille facility (9–30 m fetch) and highest in the circular facilities with quasi-unlimited fetch. The values in the large Delft facility at 90 m fetch are in between these values.

These results indicate that the energy balance in the high frequency waves (> 10 Hz) is quite a complicated process. It cannot be just a local balance between wind input, spectral flux divergence and dissipation as assumed by *Phillips* [1985] (3). Because of the fetch dependence, the large-scale waves also seem to influence the mean energy balance of the small-scale waves. This may either be due to bias effects caused by modulation of the small scales or a strong direct energy flux from large-scale waves to small-scale waves.

The spectral densities from the Hamburg facility at 15 m fetch (Fig. 5b and Fig. 8) constitute a special case. It happens that these experiments were obtained with a water surface contaminated by a surfactant of unknown origin. They demonstrate the influence of surface contamination on the wave slope spectrum. At low wind speeds, the development of the high frequency waves is strongly hindered. Then the spectral density increases very strongly and finally reaches similar values as in the other facilities with a clean(er) water surface. After the tank was emptied and carefully cleaned, one more measurement was conducted that agrees very well with the Marseille values.

5.3 Wavenumber Spectra

From the image sequences obtained with the ISG, two-dimensional wavenumber spectra have been calculated from a 0.8 m × 0.6 m large image sector. (With smaller image sectors even the smallest capillary waves (2 mm wave length) can be measured with the ISG). The two-dimensional wavenumber spectra shown in Fig. 9 have been averaged over 30 images with along-wind illumination and show wavelength from 1–50 cm. The spectral densities have been multiplied with k^2 in order to get equal degrees of saturation as constant spectral densities (see (9)). In addition, a correction factor $\cos^{-2}\Theta$ has been applied to obtain the total slope spectrum from the measured along-wind slope spectrum (compare (9)). This correction factor becomes very large for waves traveling nearly cross-wind, since their along-wind slope component is very low. Therefore a sector in cross-wind direction has been omitted where no reliable spectral densities can be obtained. (These missing values can be obtained with cross-wind illumination. Such images have been recorded, but have not been evaluated yet.)

In the spectra at the two lower wind speeds (Fig. 9a and b), a sharp decrease of the spectral densities is observed beyond wavenumbers of 180 m^{-1} ($\lambda < 3.5$ cm). This decrease corresponds to the decrease in the slope frequency spectra at about 10 Hz (Fig. 7a). At 4.1 m/s (Fig. 9b), a nearly uniform angular distribution shows up in the measurable angular sector of ±72°. At 2.2 m/s (Fig. 9b), the angular dispersion is narrower, but there is a slight tendency to a bimodal distribution for wavenumbers of about 100 m^{-1} ($\lambda = 6$ cm).

Finally, at 8 m/s wind (Fig. 9b), the spectrum first slightly increases and then decreases again, qualitatively supporting the considerations about the equilibrium range in section 3.2. More data will be available soon, which allow a quantitative analysis.

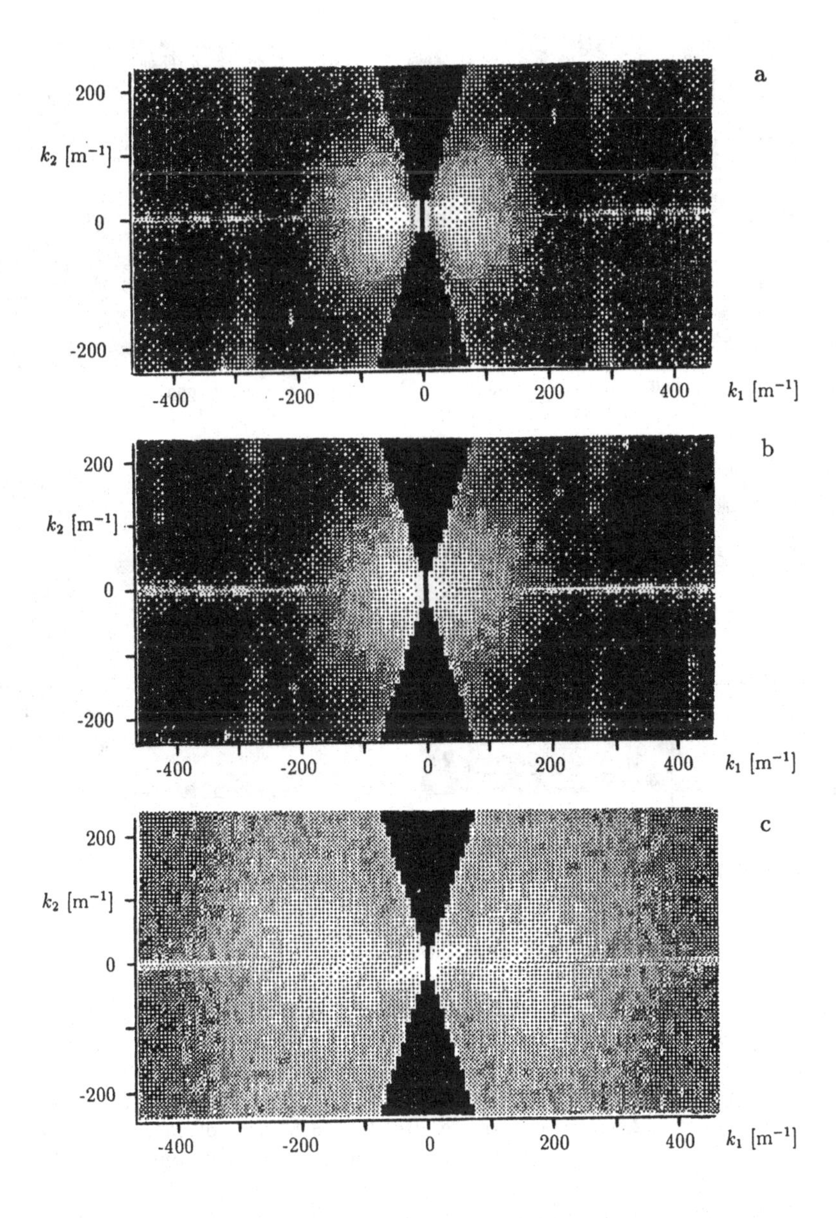

Figure 9: Wavenumber spectra calculated from the ISG images with along-wind illumination at 100 m fetch; wind speed [m/s]: a) 2.2, b) 4.1, c) 8.0; spectra multiplied with k^2; logarithmic gray scale covering 2 decades.

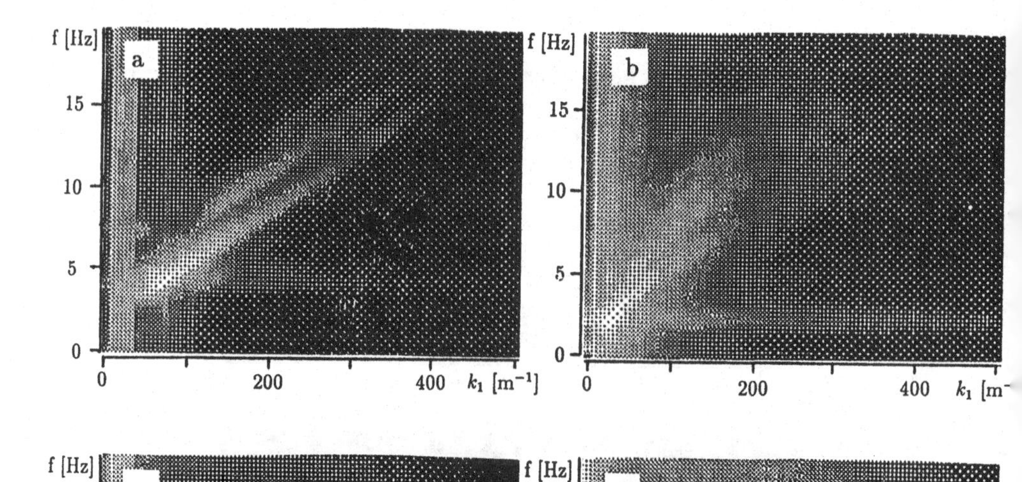

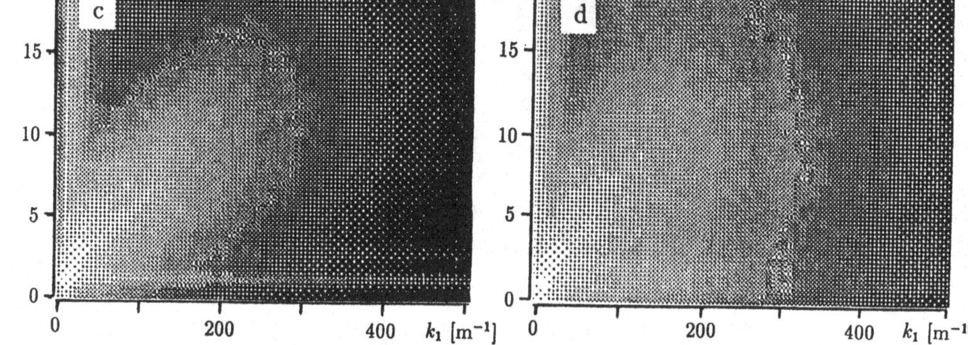

Figure 10: Wavenumber-frequency spectra calculated from ISG images sequences obtained at 40 m fetch; wind speed [m/s]: a) 1.9, b) 3.1, c) 8.0, d) 12 m/s; logarithmic gray scale covering 6 decades.

5.4 Wavenumber-Frequency Spectra

Wave image sequences give the full information about the space-time properties of the waves. In principle, it is possible to calculate three-dimensional wavenumber-frequency spectrum from the image sequences. Unfortunately, enormous amounts of data are involved with such a procedure. One image sequence of just 20 s duration (512 images with a resolution of 512 × 512 pixels) constitute already 128 Mbytes of data. To get a first hint on the potentiality of wave image sequences, we processed images containing one space and one time coordinate (space-time images).

Such images are formed taking one along-wind oriented line out of each image. An example is shown in Fig. 1c of [*Halsema et. al.*, 1988]. In such images the information on the wave direction is lost, since only k_1 is obtained. Yet k_1 is a good approximation for k. Even in the extreme case of a uniform angular distribution, the mean k_1 value is only 15% lower than k for the along-wind slope component. The spectra were averaged over 8 image sequences. They show slight disturbances (horizontal and vertical stripes), since no window has been applied in this preliminary analysis prior to the Fourier transform processing.

Two very interesting effects can be observed:

First, at the lower wind speeds (Fig. 10a and b), the strong nonlinearity of the wave field can be observed. At 1.9 m/s wind (Fig. 10a), the spectrum splits up into three parallel bands. The two additional bands constitute the first and second harmonics of the first one. The peaks of the three bands lie on a line through the origin. Consequently, they are traveling with the same speed. At the same wavenumber, there are waves with a lower frequency traveling slower and independently of the harmonics. The spectral densities of both components are about the same. Thus the data directly prove that the wave field consists about equally of free waves and harmonics of larger scale waves. This result is not surprising, since the time series of the laser slope gauge already showed the strong nonlinearity of the waves (Fig. 3a).

At higher wind speeds, the split is blurred by another effect. Now larger dominant waves are present causing a strong modulation of the phase speed and thus the frequency of the smaller scale waves. Figure 10c and d nicely show how the modulation related frequency dispersion increases with wind speed.

6 Conclusions

Though only a preliminary analysis of the data has been undertaken, a number of interesting results have been gained. On the one hand, the data show a rough agreement with the spectral shapes as given by the energy balance considerations in sections 2 and 3. Obviously, the frequency spectra are considerably disturbed by the modulation of the wave frequency due to the orbital velocities of the large scale waves. This effect may smear the spectral shapes considerably and partly explain the deviations from the expected spectral shapes.

On the other hand, significant deviations are obvious. The dependence of the spectral densities on the friction velocity is not described accurately. Considerable nonlinearities revealed by the slope data, question the whole concepts considered so far.

This paper demonstrates the great potentiality of wave slope measurements, especially the area-extended measurements with the imaging slope gauge. The new data are a challenge to improve the theoretical understanding of the energy balance in small-scale waves.

ACKNOWLEDGEMENTS. This paper covers measurements which partly have been obtained in different cooperations. The wave measurements in the IMST facility Marseille were done in two campaigns, one in 1981 together with M. Coantic (IMST) and L. Merlivat, the other during a half year research visit of the author at the IMST in 1984 in a programme together with A. Ramamonjiarisoa (IMST) and A. Lifermann (CNES, Toulouse). The measurements in the Delft wind-wave flume are part of the VIERS-1 project [*Halsema et al.*, 1988] including the participation of D. van Halsema (TNO), W. Oost (KNMI), P. Snoeij (Technical University of Delft), W. de Voogt (Delft Hydraulics), and P. Libner, K. Riemer, S. Waas, D. Wierzimok (Heidelberg University). Financial support by the German Science Foundation, the NATO Air-Sea Interaction Panel, EDF, CNES, IFREMER, and the state of Baden-Württemberg is gratefully acknowledged.

Literature

Coantic, M., Mass transfer across the ocean/air interface: small scale hydrodynamic and aerodynamic mechanisms, *Physicochemical Hydrodynamics, 1*, 249–279, 1980.

Cox, C. S., Measurements of slopes of waves, *J. Marine Res., 16*, 199–230, 1958.

Jähne, B., K. O. Münnich, R. Bösinger, A. Dutzi, W. Huber, and P. Libner, On the parameters influencing air-water gas exchange, *J. Geopys. Res, 92*, 1937–1949, 1987.

Kitaigorodskii, S. A., On the theory of the equilibrium range in the spectrum of wind-generated gravity waves, *J. Phys. Oceanogr., 13*, 816–827, 1983.

Lange, P. A., B. Jähne, J. Tschiersch, and J. Ilmberger, Comparison between an amplitude-measuring wire and a slope-measuring laser water wave gauge, *Rev. Sci. Instrum., 53*, 651–655, 1982.

Long, S. R., and N. E. Huang, On the variation and growth of wave-slope spectra in the capillary-gravity range with increasing wind, *J. Fluid Mech., 77*, 209–228, 1976.

Lubard, S. C., J. E. Krimmel, L. R. Thebaud, D. D. Evans, and O. H. Shemdin, Optical image and laser slope meter intercomparisons of high-frequency waves, *J. Geophys. Res., 85*, 4996–5002, 1980

Phillips, O. M., The equilibrium range in the spectrum of wind-generated waves, *J. Fluid Mech., 4*, 426–434, 1958.

Phillips, O. M., Spectral and statistical properties of the equilibrium range in wind-generated gravity waves, *J. Fluid Mech., 156*, 505–531, 1985.

Reece, A. M., Modulation of short waves by long waves, *Boundary Layer Meteorol., 13*, 203–214, 1978.

Rosenthal, W., Thoughts on the energy-momentum balance, this volume, 1988.

Tang, S., and O. H. Shemdin, Measurement of high frequency waves using a wave follower, *J. Geophys. Res, 88*, 9832–9840, 1983.

van Halsema, D., B. Jähne, W. A. Oost, C. Calkoen, and P. Snoeij, First results of the VIERS-1 experiment, this volume, 1988.

ON THE CHANGES IN PHASE SPEED OF ONE TRAIN OF WATER WAVES IN THE PRESENCE OF ANOTHER

S.J. Hogan, Mathematical Institute, University of Oxford, 24-29 St. Giles', Oxford OX1 3LB, U.K.

Idith Gruman, Department of Civil Engineering, Technion Haifa 32000, Israel

M. Stiassnie, School of Mathematics, University of Bristol, Bristol BS8 1TW, U.K.

ABSTRACT. We present calculations of the change in phase speed of one train of water waves in the presence of another. We use a general method, based on Zakharov's (1968) integral equation. In the important case of gravity-capillary waves, we present the correct form of the Zakharov kernel. This is used to find the expressions for the changes in phase-speed. These results are then checked using a perturbation method based on that of Longuet-Higgins and Phillips (1962). Agreement to 6 significant digits has been obtained between the calculations based on these two distinct methods. Full numerical results in the form of polar diagrams over a wide range of wavelengths, away from conditions of triad resonance, are provided.

1. INTRODUCTION

Longuet-Higgins and Phillips (1962) showed that the phase speed of one wave train on the surface of an ideal fluid is modified in the presence of another, and vice versa. These changes in phase speed can be traced to the fact that surface gravity waves interact in sets of four waves if the resonance conditions

$$\sigma_1 \pm \sigma_2 \pm \sigma_3 \pm \sigma_4 = 0 \tag{1.1}$$

$$\underline{k}_1 \pm \underline{k}_2 \pm \underline{k}_3 \pm \underline{k}_4 = 0 \tag{1.2}$$

are met. Here σ is the linearized wave frequency and $\underline{k}$ the wavenumber related by the dispersion relation

$$\sigma_i^2 = g|\underline{k}_i| \qquad (i = 1,2,3,4) \tag{1.3}$$

where g is the acceleration due to gravity.

G. J. Komen and W. A. Oost (eds.), Radar Scattering from Modulated Wind Waves, 121–137.

The present study generalizes this earlier work in a number of important ways. In Section 2, we give a general method to find the change in phase speed. This is based on Zakharov's (1968) equation and shows that we only need the Zakharov kernel $T(\underline{k},\underline{k}_1,\underline{k}_2,\underline{k}_3)$ in order to obtain this change. In Section 3, we compare our results in the case of gravity-only propagation with those of Longuet-Higgins and Phillips (1962). We find full agreement with their work, provided a small printing error is corrected. The important case of gravity-capillary waves is then considered. We find that the kernel $T(\underline{k},\underline{k}_1,\underline{k}_2,\underline{k}_3)$ given by Zakharov (1968) is incorrect. We give the correct form in Section 4. In Section 5 we give an alternative derivation of Δc_2 in the case of gravity-capillary waves, based on the perturbation analysis of Longuet-Higgins and Phillips (1962). The two separate expressions for Δc_2 are found to yield identical numerical values. We give some simple expressions for Δc_2 in Section 6 for the case of collinear propagation. In Section 7, we give full numerical results in the form of polar diagrams over a wide range of wavelengths, for arbitrary angle of intersection of the wavetrains, and discuss their significance. Regions of possible triad interaction are indicated. Section 8 is devoted to a summary of our results.

2. GENERAL METHOD

Zakharov's (1968) integral equation is given by

$$i \frac{\partial B}{\partial t}(\underline{k},t) = \iiint_{-\infty}^{\infty} T(\underline{k},\underline{k}_1,\underline{k}_2,\underline{k}_3)\, B^*(\underline{k}_1,t)\, B(\underline{k}_2,t)\, B(\underline{k}_3,t)$$
$$\times\, \delta(\underline{k} + \underline{k}_1 - \underline{k}_2 - \underline{k}_3)$$
$$\times\, \exp\{i[\sigma(\underline{k}) + \sigma(\underline{k}_1) - \sigma(\underline{k}_2) - \sigma(\underline{k}_3)]\}d\underline{k}_1 d\underline{k}_2 d\underline{k}_3 \tag{2.1}$$

where $B(\underline{k},t)$ is related to the free surface $\zeta(\underline{x},t)$ by

$$\zeta(\underline{x},t) = \frac{1}{2\pi}\int_{-\infty}^{\infty} \left(\frac{|\underline{k}|}{2\sigma(\underline{k})}\right)^{\frac{1}{2}} \{B(\underline{k},t)\exp\{i(\underline{k}.\underline{x} - \sigma(\underline{k})t)\} + *\}d\underline{k} \tag{2.2}$$

The complex conjugate is denoted by $*$, $\underline{k}$ is the wave vector, $\underline{x}$ is the horizontal spatial vector and t is time. The linearized wave frequency σ is related to $\underline{k}$ through the linear dispersion relation of the waves of interest. The kernel $T(\underline{k},\underline{k}_1,\underline{k}_2,\underline{k}_3)$ is a real function of its variables, and is taken in the form symmetric in $\underline{k}_2,\underline{k}_3$ which is uniquely defined (see section 3 in Stiassnie & Shemer (1984). For strict resonance conditions, given by Equations (1.1), (1.2), T is also symmetric in its first two arguments $\underline{k}$, $\underline{k}_1$.

We now consider the consequence of taking two weakly nonlinear wave trains, denoted by 1 and 2, to make up the free surface and look for simple phase-change solutions that might result. Thus we take

$$B(\underline{k},t) = B_1(t)\, \delta(\underline{k}-\underline{k}_1) + B_2(t)\, \delta(\underline{k}-\underline{k}_2) \tag{2.3}$$

and substitute equation (2.3) into eqution (2.1) to find

$$i \frac{dB_1}{dt} = T(\underline{k}_1,\underline{k}_1,\underline{k}_1,\underline{k}_1)|B_1|^2 B_1 + \{T(\underline{k}_1,\underline{k}_2,\underline{k}_1,\underline{k}_2) + T(\underline{k}_1,\underline{k}_2,\underline{k}_2,\underline{k}_1)\}|B_2|^2 B_1 \tag{2.4}$$

and

$$i \frac{dB_2}{dt} = T(\underline{k}_2,\underline{k}_2,\underline{k}_2,\underline{k}_2)|B_2|^2 B_2 + \{T(\underline{k}_2,\underline{k}_1,\underline{k}_2,\underline{k}_1) + T(\underline{k}_2,\underline{k}_1,\underline{k}_1,\underline{k}_2)\}|B_1|^2 B_2$$

We denote $T(\underline{k}_1,\underline{k}_1,\underline{k}_1,\underline{k}_1)$ by T_1 and $T(\underline{k}_2,\underline{k}_2,\underline{k}_2,\underline{k}_2)$ by T_2. The above mentioned symmetry properties of $T(\underline{k},\underline{k}_1,\underline{k}_2,\underline{k}_3)$, allow us to write both of the expressions in the curly brackets of equations (2.4) and (2.5), as $2\,T(\underline{k}_1,\underline{k}_2,\underline{k}_1,\underline{k}_2)$ which we will denote by $2T_{1,2}$. We also denote $|\underline{k}_i|$ as k_i and $\sigma(\underline{k}_i)$ as σ_i.

The solution of (2.4) and (2.5) is given by

$$B_1(t) = A_1 \exp\{-i(T_1A_1^2 + 2T_{1,2}A_2^2)t\} \tag{2.6}$$

$$B_2(t) = A_2 \exp\{-i(T_2A_2^2 + 2T_{1,2}A_1^2)t\} \tag{2.7}$$

where A_1 and A_2 are constants.

We now substitute equations (2.3), (2.6) and (2.7) into equation (2.2). We can write the result in the form

$$\zeta(\underline{x},t) = a_1 \cos(\underline{k}_1.\underline{x} - \Omega_1 t) + a_2 \cos(\underline{k}_2.\underline{x} - \Omega_2 t) \tag{2.8}$$

where

$$A_i = 2\pi \left(\frac{\sigma_i}{2k_i}\right)^{\frac{1}{2}} a_i \qquad i = 1,2 \tag{2.9}$$

The frequencies of the wave trains are given by

$$\Omega_1 = \sigma_1 + T_1A_1^2 + 2T_{1,2}A_2^2 \tag{2.10}$$

and

$$\Omega_2 = \sigma_2 + T_2A_2^2 + 2T_{1,2}A_1^2 \tag{2.11}$$

The change in the frequency of each wave train is therefore made up of two parts. In equation (2.11) for example, the first correction to σ_2 is given by $T_2A_2^2$ which is the well-known Stokes (1847) correction. This term is due to the nonlinearity of the wave train itself and is present even if the other wave train is absent. The second correction is given by $2T_{1,2}A_1^2$ and is entirely due to the presence of the other wave train. It is the same order as the usual Stokes correction.

The phase speed of the wave train can be given by

$$c_2 = \Omega_2/k_2 \qquad (2.12)$$

and so the change in phase speed of the weakly nonlinear wave train 2 due to the presence of wave train 1, Δc_2, is given by

$$\Delta c_2 = c_2 - \frac{(\sigma_2 + T_2 A_2^{\ 2})}{{}_2k} \qquad (2.13)$$

$$= \frac{2T_{1,2} A_1^{\ 2}}{k_2} \qquad (2.14)$$

Using equation (2.9), this becomes

$$\Delta c_2 = \frac{4\pi^2 \sigma_1}{k_1 k_2} T_{1,2}\, a_1^{\ 2} \qquad (2.15)$$

Thus we only need $T_{1,2}$ to evaluate Δc_2 and we have shown that this change in phase velocity is independent of the amplitude of wave train 2. This latter result was obtained for gravity waves only by Longuet-Higgins and Phillips (1962). We see now that it is true in general.

In a similar way one can show that the change in phase speed of the weakly nonlinear wave train 1 due to the presence of wave train 2 is

$$\Delta c_1 = \frac{4\pi^2 \sigma_2}{k_1 k_2} T_{1,2}\, a_2^2 \qquad (2.16)$$

3. GRAVITY WAVES

In the case of gravity waves propagating on the surface of an ideal fluid of infinite depth, the linearised dispersion relation is given by equation (1.3). The Zakharov kernel function $T(\underline{k},\underline{k}_1,\underline{k}_2,\underline{k}_3)$ is readily available in compact form in Crawford et al (1981).

Longuet-Higgins and Phillips (1962) obtained the expression

$$\Delta c_2 = K'/2a_2\sigma_2^{\ 2} \qquad (3.1)$$

where K' is given in their equation (2.8). There is a misprint in their expression for K' (see Hogan, Gruman, Strassnie, 1988).

We have obtained numerical agreement between equation (2.15) and equation (3.1) for arbitrary values of θ (defined in Figure 1).

4. THE FUNCTION $T(k,k_1k_2,k_3)$ FOR GRAVITY-CAPILLARY WAVES

The linearised dispersion relation in this case is given by

$$\sigma^2 = gk + S\,k^3 \tag{4.1}$$

where S is the surface tension coefficient divided by the density of the fluid. The kernel $T(\underline{k}_0,\underline{k}_1,\underline{k}_2,\underline{k}_3)$ was given originally by Zakharov (1968). In fact the third order interaction coefficient $W_{0,1,2,3}$ (using the notation of Crawford et al (1981)) should be modified to become

$$W'_{0,1,2,3} = W_{0,1,2,3} - \frac{S}{32\pi^2}\frac{(k_0k_1k_2k_3)^{\frac{1}{2}}}{(\sigma_0\sigma_1\sigma_2\sigma_3)^{\frac{1}{2}}}\,[(\underline{k}_0\cdot\underline{k}_1)(\underline{k}_2\cdot\underline{k}_3) + (\underline{k}_0\cdot\underline{k}_2)(\underline{k}_1\cdot\underline{k}_3) + (\underline{k}_0\cdot\underline{k}_3)(\underline{k}_1\cdot\underline{k}_2)] \tag{4.2}$$

We obtained equation (4.2), following the procedure given in section VIA of Yuen and Lake (1982).

5. PERTURBATION METHOD

In this section we summarise an independent method for deriving Δc_2 based on the work of Longuet-Higgins and Phillips (1962). Let

$$A = \frac{a_1\,a_2}{2[(\sigma_1-\sigma_2)^2 - (g+S|\underline{k}_1-\underline{k}_2|^2)|\underline{k}_1-\underline{k}_2|]} \times \{[\sigma\,k_1 - \sigma_2k_2 + (\sigma_2k_1-\sigma_1k_2)\cos\theta](g + S|\underline{k}_1-\underline{k}_2|^2) - (\sigma_1-\sigma_2)[\sigma_1^2 + \sigma_2^2 - 2\,\sigma_1\sigma_2\cos^2\tfrac{1}{2}\theta]\} \tag{5.1}$$

$$B = \frac{a_1\,a_2}{2[(\sigma_1+\sigma_2)^2 - (g+S|\underline{k}_1+\underline{k}_2|^2)\,|\underline{k}_1+\underline{k}_2|]} \times \{-[\sigma_1k_1 + \sigma_2k_2 + (\sigma_1k_2 + \sigma_2k_1)\cos\theta]\,(g+S|\underline{k}_1+\underline{k}_2|^2) + (\sigma_1+\sigma_1)[\sigma_1^2 + \sigma_2^2 + 2\,\sigma_1\sigma_2\sin^2\tfrac{1}{2}\theta]\} \tag{5.2}$$

$$C = \frac{a_1\,a_2}{2[(\sigma_1-\sigma_2)^2 - (g+S|\underline{k}_1-\underline{k}_2|^2)\,|\underline{k}_1-\underline{k}_2|]} \times \{[\sigma_1k_1 - \sigma_2k_2 + (\sigma_2k_1 - \sigma_1k_2)\cos\theta]\,(\sigma_1-\sigma_2) - |\underline{k}_1-\underline{k}_2|\,[\sigma_1^2 + \sigma_2^2 - 2\,\sigma_1\sigma_2\cos^2\tfrac{1}{2}\theta]\} \tag{5.3}$$

$$D = \frac{a_1 \ a_2}{2[(\sigma_1+\sigma_2)^2 - (g+S|\underline{k}_1+\underline{k}_2|^2) \ |\underline{k}_1+\underline{k}_2|]}$$

$$\times \{-[\sigma_1 k_1 + \sigma_2 k_2 + (\sigma_1 k_2 + \sigma_2 k_1) \cos \theta] \ (\sigma_1+\sigma_2)$$

$$+ |\underline{k}_1+\underline{k}_2| \ [\sigma_1^2 + \sigma_2^2 + 2 \ \sigma_1\sigma_2 \sin^2 \tfrac{1}{2} \theta]\} \tag{5.4}$$

Thus the change in the phase velocity of wave 2 in the presence of wave 1 is given by

$$\Delta \ c_2 = \frac{\sigma_2}{k_2} \cdot \frac{\delta}{a_2\sigma_2} = \frac{\delta}{a_2 k_2} \tag{5.5}$$

where the coefficient δ is given by the expression

$$\delta = a_1/4\sigma_2 \ \{A|\underline{k}_1-\underline{k}_2| \ [\sigma_2 k_2 - \sigma_2|\underline{k}_1-\underline{k}_2| - \cos \alpha \ (\sigma_2 \ k_1 + \sigma_1 \ k_2)]$$

$$+ B \ |\underline{k}_1+\underline{k}_2| \ [\sigma_2 k_2 - \sigma_2|\underline{k}_1+\underline{k}_2| - \cos \beta \ (\sigma_2 k_1 - \sigma_1 k_2)]$$

$$+ \sigma_1 k_2(\sigma_2\cos\theta-\sigma_1)(C-D)$$

$$+ a_1 a_2 k_2[\sigma_1{}^2 k_1 + \sigma_1\sigma_2(2 \ k_1 + k_2) \cos \theta$$

$$- 1/2 \ S \ k_1{}^2 k_2{}^2(\sin^2\theta + 3 \cos^2\theta)]\} \tag{5.6}$$

where A, B, C and D are given in equations (5.1)-(5.4) and the angles α,β and θ are defined in Figure 1.

We have also shown numerically that equation (5.5) and (5.6) does equal equation (2.15) when the modification suggested in equation (4.2) is applied. Full details are given in Hogan et al. 1988.

6. COLLINEAR WAVE TRAINS

We now evaluate the change in phase velocity in four special cases. Thus we consider $\underline{k}_1$ parallel and antiparallel to $\underline{k}_2$ with $k_1 > k_2$ and $k_1 < k_2$. We consider flow under gravity only, under surface tension only and under the combined effects of gravity and surface tension.

(a) Gravity only

We set S = 0 in equation (5.5) For $\underline{k}_1$ parallel to $\underline{k}_2$ and $k_1 < k_2$ (case (i)) we set $\theta = \alpha = 0$ and $\beta = \pi$ in Figure 1 and find

$$\Delta \ c_2 = a_1^2 \ k_1 \ \sigma_1 \tag{6.1}$$

If $k_1 > k_2$, (case (ii)), then $\theta = 0$, $\alpha = \beta = \pi$ and we find

$$\Delta c_2 = a_1^2 k_2 \sigma_1 \tag{6.2}$$

For $\underline{k}_1$ parallel to $-\underline{k}_2$ and $k_1 < k_2$ (case (iii)) we take $\theta = \alpha = \pi$ and $\beta = 0$ to find

$$\Delta c_2 = -a_1^2 k_1 \sigma_1 \tag{6.3}$$

and if $k_1 > k_2$ (case (iv)) we take $\theta = \alpha = \beta = \pi$ and find

$$\Delta c_2 = -a_1^2 k_2 \sigma_1 \tag{6.4}$$

In this way we recover the results in Section 3 of Longuet-Higgins and Phillips (1962).

(b) Surface tension only

Here we set $g = 0$ and take $\sigma_i^2 = Sk_i^3$ $(i = 1,2)$. Thus for case (i) we find

$$\Delta c_2 = \frac{1}{8} (a_1 k_1)^2 \frac{\sigma_2}{k_2} H(\eta) \tag{6.5}$$

where $\eta^2 = k_1/k_2$

$$H(\eta) = \frac{-27 + 72\eta - 78\eta^2 + 32\eta^3 + 27\eta^4 - 24\eta^5 - 32\eta^6 + 48\eta^7 - 18\eta^8}{(9 - 12\eta + 4\eta^2 - 3\eta^4 - 4\eta^5 + 6\eta^6)} \tag{6.6}$$

For case (ii), we obtain

$$\Delta c_2 = \frac{1}{8} (a_1 k_1)^2 \frac{\sigma_2}{k_2} H(\frac{1}{\eta}) \tag{6.7}$$

For the two antiparallel cases, the results are similar (but not the negative of equations (6.5) and (6.7)). In case (iii), we find

$$\Delta c_2 = \frac{1}{8} (a_1 k_1)^2 \frac{\sigma_2}{k_2} H(-\eta) \tag{6.8}$$

and for case (iv),

$$\Delta c_2 = \frac{1}{8} (a_1 k_1)^2 \frac{\sigma_2}{k_2} H(-1/\eta) \tag{6.9}$$

It is clear that we need only consider the value of $H(\eta)$ for $-1 \leqslant \eta \leqslant 1$ in order to cover all four cases above. This is done in Figure 2 where we plot $H(\eta)$ in the range $-1 \leqslant \eta \leqslant 1$. Note that $H(\eta)$ is always negative and monotonically increases from $H = -8$ at $\eta = -1$ to $H = -1$ at $\eta = +1$.

Thus we find that whereas for gravity - only motion parallel propagation leads to an increase and antiparallel propagation to a decrease in phase speed, the quantity Δc_2 is always negative for surface tension only motion.

We note that the expression for $H(\eta)$ has been successfully used to verify computer calculations of the normal mode perturbations to fully nonlinear pure capillary waves (Hogan 1988).

(c) Gravity and surface tension

The various expressions for Δc_2 in the combined case of gravity and surface tension propagation are now derived for the four special cases. Let us take

$$R_j = \frac{g}{Sk_j^2} \ (j = 1,2) \text{ and } r = \frac{S\, k_1 k_2}{4\, \sigma_1 \sigma_2} \tag{6.10}$$

where now $\sigma_j^2 = g\, k_j + Sk_j^3 = Sk_j^3 \ (1 + R_j) \qquad (j = 1,2)$.

Then we find for case (i)

$$\frac{\Delta c_2}{(\Delta c_2)_{S=0}} = \frac{1}{2}\, r \left(\frac{I\,(k_1,k_2) + r\, J\,(k_1,k_2)}{K\,(k_1,k_2) - r\, L\,(k_1,k_2)}\right) = F(k_1,k_2) \tag{6.11}$$

where $(\Delta c_2)_{S=0}$ is the value of $\Delta\, c_2$ in the relevant gravity - only case evaluated at the same frequency $\sigma_2 = \sigma_2(S=0)$. For case (i) here this is given by equation (6.1). The functions $I\,(k_1,k_2)$, $J\,(k_1,k_2)$, $K\,(k_1,k_2)$ and $L\,(k_1,k_2)$ are given by

$$I\,(k_1,k_2) = 9k_2(k_1^2-2k_2^2) - 3k_1^2(4k_1+k_2)(1+R_1) - 8k_1k_2^2(1+R_1)(1+R_2) \tag{6.12}$$

$$J\,(k_1,k_2) = 27\, k_2^2(k_2^2-k_1^2) + 18\, k_1(k_1^3+k_2^3)(1+R_1)$$
$$+ 32\, k_1^3k_2(1+R_1)^2(1+R_2) + 60\, k_1k_2^3(1+R_1)(1+R_2) \tag{6.13}$$

$$K\,(k_1,k_2) = k_1^2(1+R_1) + 3\, k_2^2 \tag{6.14}$$

$$L\,(k_1,k_2) = 9k_2(k_2^2-k_1^2) + 6k_1^2(k_1+k_2)(1+R_1) + 4k_1k_2^2(1+R_1)(1+R_2) \tag{6.15}$$

For case (ii), the result by symmetry,

$$\frac{\Delta c_2}{(\Delta c_2)_{S=0}} = F(k_2,k_1) \tag{6.16}$$

where $(\Delta\, c_2)_{S=0}$ is now given in equation (6.2).

For antiparallel propagation we find for case (iii) that

$$\frac{\Delta c_2}{(\Delta c_2)_{S=0}} = \frac{1}{2}\, r \left\{ \frac{-I\,(k_1,k_2) + r\, J\,(k_1,k_2)}{K\,(k_1,k_2) + r\, L\,(k_1,k_2)}\right\} = G\,(k_1,k_2) \tag{6.17}$$

where $(\Delta\, c_2)_{S=0}$ is given in equation (6.3) and for case (iv), by symmetry,

$$\frac{\Delta c_2}{(\Delta c_2)_{S=0}} = G(k_2,k_1) \tag{6.18}$$

where $(\Delta\, c_2)_{S=0}$ is given in equation (6.4).

Finally in this section we point out that it is not always possible to obtain a finite value for $\Delta c_2/(\Delta c_2)_{S=0}$ because of triad interactions. Energy is continually transferred between the modes and so no steady solutions exist in general at those values of the resonant wavenumbers. This situation is unavoidable and limits the application of our results.

7. RESULTS AND DISCUSSION

7.1. Results

The analytic formulae of the previous section are now supplemented by numerical results for realistic cases with arbitrary angle of intersection θ to guide our understanding.

We construct the quantity $\tilde{T}_{1,2}$ given by

$$\tilde{T}_{1,2} = \frac{4\pi^2 T_{1,2}}{k_1 k_2^2} = \frac{\delta}{\sigma_1 a_1^2 a_2 k_2^2} = \frac{\Delta c_2}{a_1^2 k_2 \sigma_1} = \frac{\Delta c_1}{a_2^2 k_2 \sigma_2} \tag{7.1}$$

This is a dimensionless function of the three parameters $\nu = k_2/k_1$, $\cos\theta = (\underline{k}_1 . \underline{k}_2)/k_1 k_2$ and R_1 where $0 < \nu \leqslant 1$ and $0 \leqslant \theta \leqslant \pi$. We present our results in the form of polar diagrams showing lines of equal $\tilde{T}_{1,2}$ for four wavelengths $\lambda_1 = 200$cm, 5cm, 1cm and 0.1cm. We take the two wavenumber vectors to be

$$\underline{k}_1 = k_1(1,0)$$

$$\underline{k}_2 = k_1(X,Y) \tag{7.2}$$

and we set $S = 74\ \text{cm}^3/\text{s}^2$

The result for $\lambda_1 = 200$cm are shown in Figure 3a and 3b. Figure 3a covers the range $0.1k_1 \leqslant k_2 \leqslant k_1$, and Figure 3b cover $0.01k_1 \leqslant k_2 \leqslant 0.1k_1$. The same arrangement holds in all other following figures. Since $\lambda_1 = 200$cm is essentially a gravity wave and since $\lambda_2 > \lambda_1$ these two diagrams summarize the results for 'pure' gravity waves. Figures 4a and 4b for $\lambda_1 = 5$cm (and again with $\lambda_2 > \lambda_1$) are for cases dominated by gravity, but somewhat influenced by surface tension. The lines of constant $\tilde{T}_{1,2}$ are rotated more to the left than in Figures 3a and 3b. This implies an increase in Δc_i when $\theta = 0$ and a decrease when $\theta = \pi$. The case $\lambda_1 = 1$cm, shown in Figures 5a and 5b, is typical for situations where gravity and surface tension have a comparable role. Here the picture is considerably different from the previous figures. The shaded area for $X,Y \geqslant 0$ represents the area where significant triad interaction takes place. This area is located around the triad resonance curves and is characterized by large values and gradients of $\tilde{T}_{1,2}$. In this area our quartet interaction solutions are not applicable, as was mentioned in section 6. We note the fact that even some cases of parallel propagation are resonant, whereas every case in the left hand side of Figures 5a, 5b is non-resonant. The final case in this series is for a 'pure' capillary wave, $\lambda_1 = 0.1$cm, in the presence of another wave. This other wave is in the capillary to gravity-capillary range, $\lambda_2 = 0.1$-1cm in Figure 6a to $\lambda_2 = 1$-10cm in Figure 6b.

Now the triad interaction area has moved off the parallel axis, in keeping with the fact that pure capillary waves resonate in non-parallel triads (McGoldrick 1965). In fact $\Delta c_1 < 0$ is the case, for almost all non-resonant values of θ.

7.2. Discussion

The change in the phase speed of one wave train due to the presence of the other, divided by its own linearised speed is given by

$$\frac{\Delta c_1}{(\sigma_1/k_1)} = \tilde{T}_{1,2} \frac{\sigma_2}{\sigma_1} \frac{k_1}{k_2} (k_2 a_2)^2 \tag{7.3}$$

$$\frac{\Delta c_2}{(\sigma_2/k_2)} = \tilde{T}_{1,2} \frac{\sigma_1}{\sigma_2} \left(\frac{k_2}{k_1}\right)^2 (k_1 a_1)^2 \tag{7.4}$$

In the sequel we focus on the effect the longer wave (No. 2) has on the shorter one (No. 1), which is more profound than the other way round. Substituting (4.1) into (7.3) yields with $K_1 = 1/R_1$

$$\frac{\Delta c_1}{(\sigma_1/k_1)} = \tilde{T}_{1,2} \left(\frac{k_1}{k_2}\right)^{1/2} \left(\frac{1 + K_1 (k_2/k_1)^2}{1 + K_1}\right)^{1/2} (k_2 a_2)^2 \tag{7.5}$$

From Figures 3 to 6 we see that the order of magnitude of $T_{1,2}$ is unity. Assuming that $(k_2 a_2)^2$ is limited by say 0.1, $\Delta c_1/(\sigma_1/k_1)$ has an upper bound of the order of 0.1μ where

$$\mu = \left(\frac{k_1}{k_2} \cdot \frac{1 + K_1 (k_2/k_1)^2}{1 + K_1}\right)^{1/2} \tag{7.6}$$

Some values of μ, which depends on two parameters K_1 and (k_1/k_2) are given in the following table.

TABLE 1. Numerical values of μ, Eq. (7.6).

k_1/k_2	$\lambda_1 = 200$cm	5cm	1cm	0.1cm
10	3.16	2.99	2.00	0.36
100	10	9.45	5.00	0.59

Thus one can see that for $k_1/k_2 = 100$, $\Delta c_1/(\sigma_1/k_1)$ can be as large as 1. Note that taking $k_1/k_2 = 100$ means that we assume the existence of a uniform wave-train (No. 1) which is several hundred wave lengths long. The realization of such circumstances is doubtful since strong modulations will probably enter and change the picture substantially.

7.3. Application

We conclude this section with a practical example. Namely the two-dimensional scattering of gravity waves, as occurs when a two-dimensional obstacle is placed in a wave-flume. For this case it is usually assumed that the reflected wave has the same frequncy as that of the incident wave. Thus $\Omega_1 = \Omega_2$, (see equations 2.10, 2.11), and one can show that the reflected wave-number $\underline{k}^{(R)}$ is related to the incident wave number $\underline{k}^{(I)}$ through

$$\underline{k}^{(R)} \approx -[1+3(k^{(I)}a^{(I)})^2(1-R_c^2)]\underline{k}^{(I)} \tag{7.7}$$

where R_c is the reflection coefficient defined by the reflected wave amplitude $a^{(R)}$ to incident wave amplitude $a^{(I)}$ ratio. From (7.7) we see that the length of the reflected wave can be significantly shorter than that of the incident wave, excluding the case of total reflection.

We can generalise equation (7.7) to include capillarity. In fact

$$k^{(R)} \approx -\{1 + \frac{[24+63K-54K^2-120K^3]}{8(1+3K)(1+4K)(1-2K)} (k^{(I)}a^{(I)})^2(1-R_c^2)\}\underline{k}^{(I)} \tag{7.8}$$

where now $K = Sk^{(I)^2}/g$. The singularity at $K = 1/2$ corresponds to subharmonic resonance, a special case of triad interaction which we have already excluded from our analysis. When $K = 0$, that is for gravity waves we recover equation (7.7). For $0.5 < K < 0.702529$, $|\underline{k}^{(R)}| < |\underline{k}^{(I)}|$ and hence the reflected wave can be longer than the incident wave, with equality at $K = 0.702529$. For capillary waves, we take K infinite and find

$$\underline{k}^{(R)} \approx -[1 + 5/8\ (k^{(I)}a^{(I)})^2(1-R_c^2)]\underline{k}^{(I)} \tag{7.9}$$

8. SUMMARY

We have shown that the Zakharov (1968) kernel function $T(\underline{k},\underline{k}_1,\underline{k}_2,\underline{k}_3)$ can be used to calculate the change in phase speed of one train of water waves in the presence of another.

For gravity-capillary waves, we have derived the correct form of $T(\underline{k},\underline{k}_1,\underline{k}_2,\underline{k}_3)$. The subsequent expression for Δc_2 has been checked against that resulting from a derivation based on the method of Longuet-Higgins and Phillips (1962). The agreement that is found gives confidence in the accuracy of both derivations.

In general our results show that surface tension effects of sufficient size change the sign as well as the magnitudes of Δc_1 and Δc_2, away from conditions of triad resonance. Some of these changes are significant enough to be measured experimentally.

ACKNOWLEDGEMENTS

S.J.H. is CEGB Research Fellow in Applied Mathematics at St. Catherine's College, Oxford. He completed part of this work as a Visiting Fellow in the School of Mathematics at the University of New South Wales. He is grateful to the Head and staff of the School for their hospitality.

Some parts of this work are based on the M.Sc. thesis of I.G. submitted to the Technion, Israel Institute of Technology.

M.S. is grateful to Professor D.H. Peregrine and the staff of the School of Mathematics, University of Bristol, for their hospitality.

REFERENCES

Crawford, D.R., Lake, B.M., Saffman, P.G. & Yuen, H.C. 1981 Stability of weakly nonlinear deep water waves in two and three dimensions. J. Fluid Mech. 105, 177-191

Hogan, S.J. 1988 The superharmonic normal mode instabilities of non-linear deep water capillary waves. J. Fluid Mech. 190, 165-177.

Hogan, S.J., Gruman, I., and Stiassnie, M. 1988 On the changes in phase speed of one train of water waves in the presence of another. J. Fluid Mech. 192, 97-114.

Longuet-Higgins, M.S. & Phillips, O.M. 1962 Phase velocity effects in tertiary wave interactions. J. Fluid Mech. 12, 333-336

McGoldrick, L.F. 1965 Resonant interactions among capillary-gravity waves. J. Fluid Mech. 21, 305-331

Stiassnie, M. and Shemer, L. 1984 On modification of the Zakharov equation for surface gravity waves. J. Fluid Mech. 143, 47-67

Stokes, G.G. 1847 On the theory of oscillatory waves. Trans. Camb. Phil. Soc. 8, 441-455

Yuen, H.C. and Lake, B.M. 1982 Nonlinear dynamics of deep-water gravity waves. Adv. Appl. Mech. 22 67-229

Zakharov, V.E. 1968 Stability of periodic waves of finite amplitude on the surface of a deep fluid. J. appl. Mech. tech. Phys. 2, 190-194

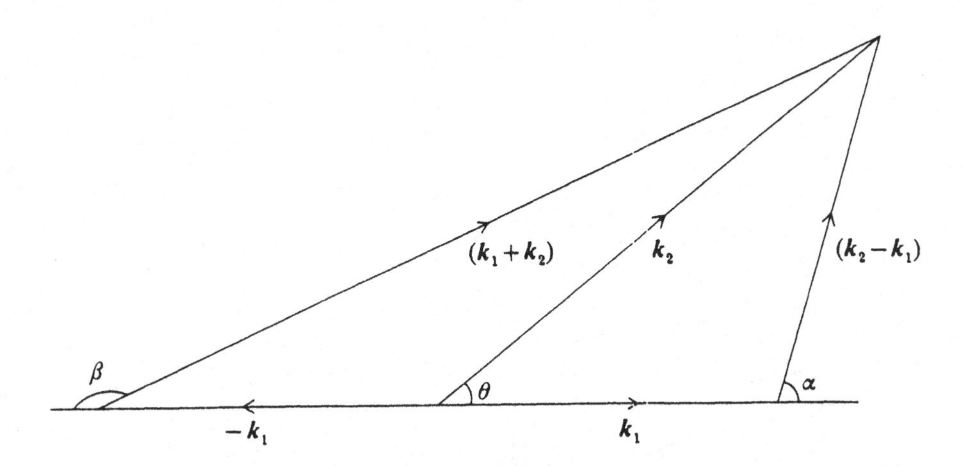

Figure 1 Definition sketch for the angles α, β and θ

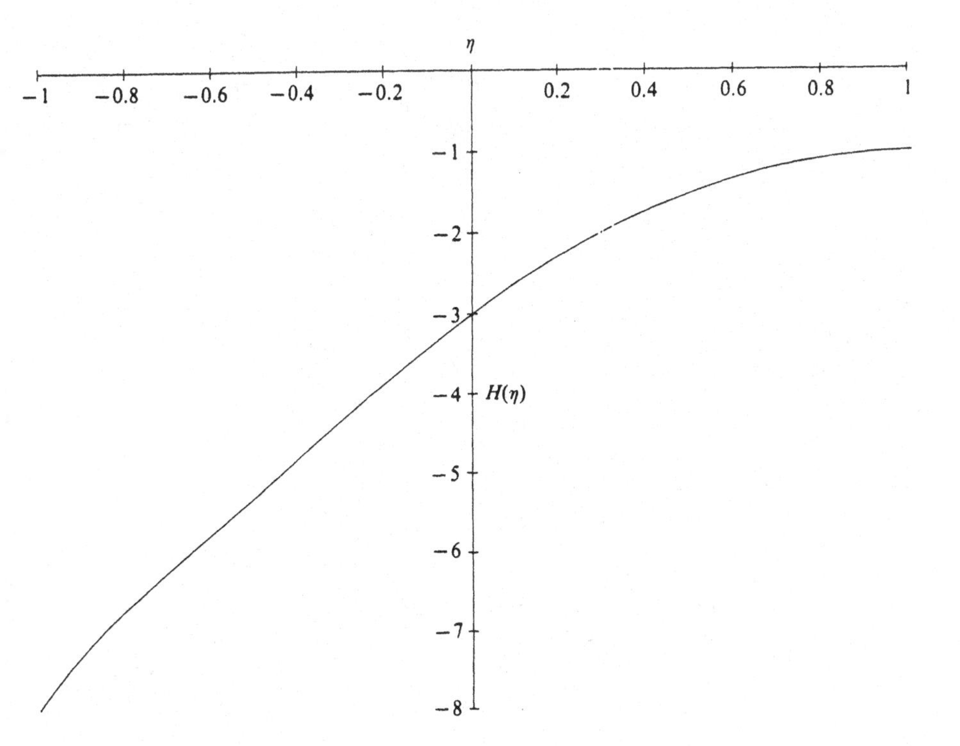

Figure 2 The function $H(\eta)$ as defined by equation (6.6) plotted in the range $-1 \leqslant \eta \leqslant 1$ where $\eta^2 = k_1/k_2$

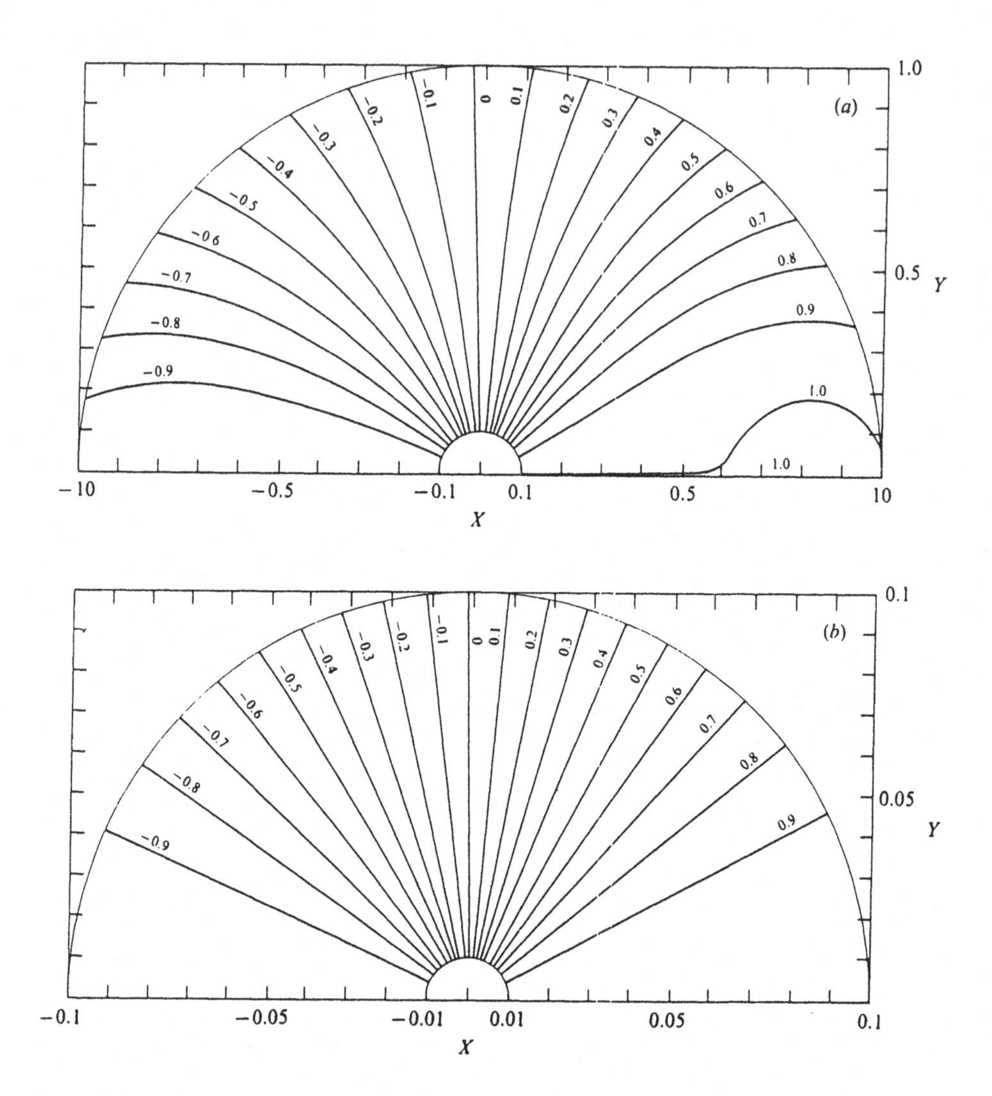

Figure 3 Lines of constant $\tilde{T}_{1,2}$ as defined in equation (7.1) for $\lambda_1 = 200$cm in the ranges: (a) $0.1\ k_1 \leqslant k_2 \leqslant k_1$, (b) $0.01k_1 \leqslant k_2 \leqslant 0.1k_1$, where $\underline{k}_1 = k_1(1,0)$ and $\underline{k}_2 = k_1(X,Y)$

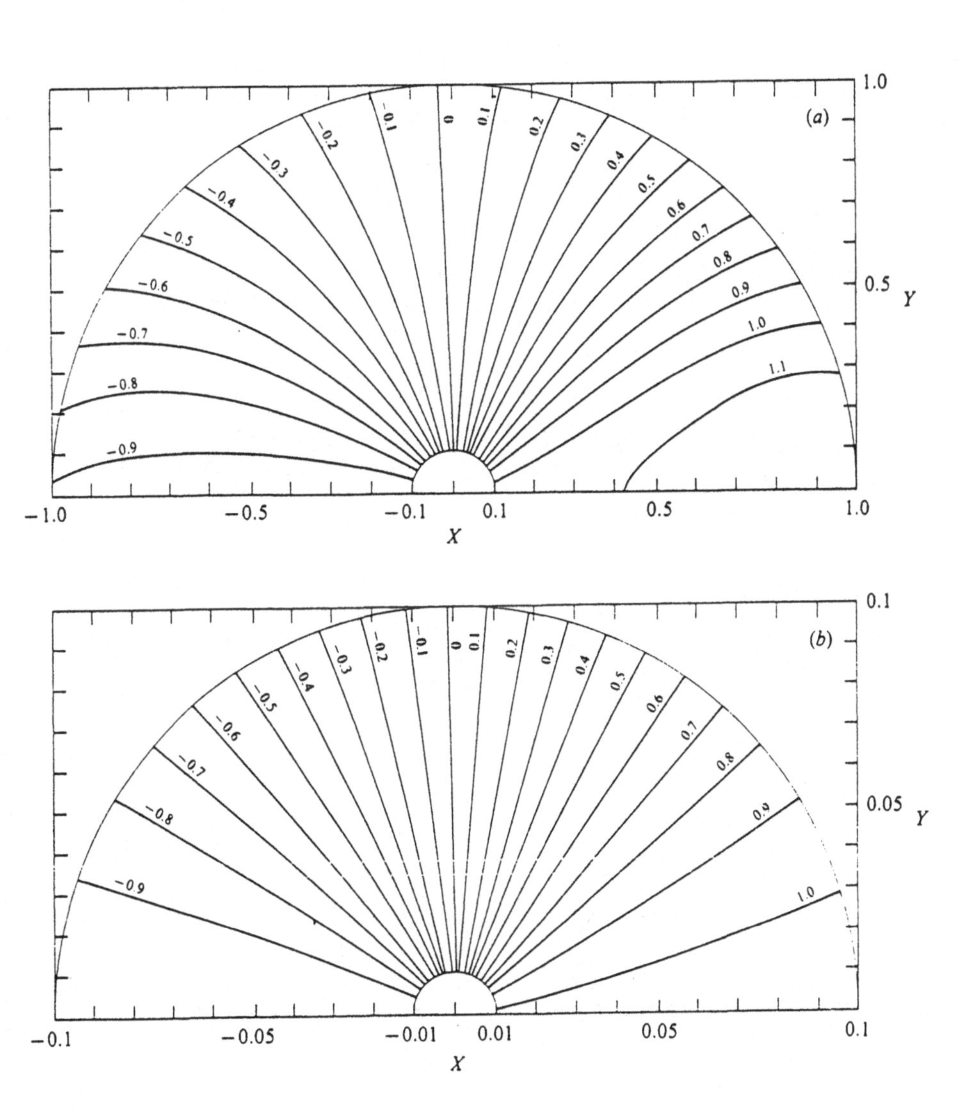

Figure 4 As in Fig. 3 for λ_1 = 5cm

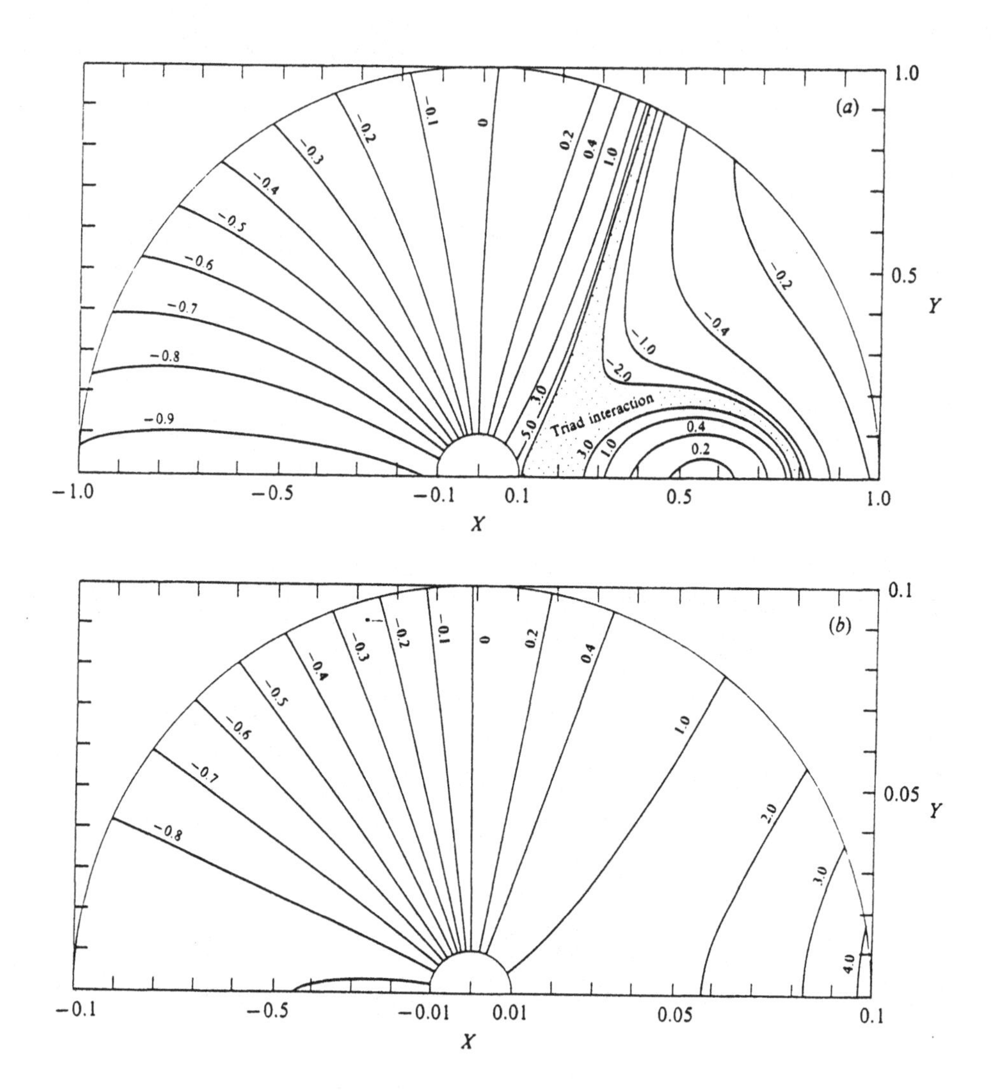

Figure 5 As in Fig. 3 for λ_1 = 1cm

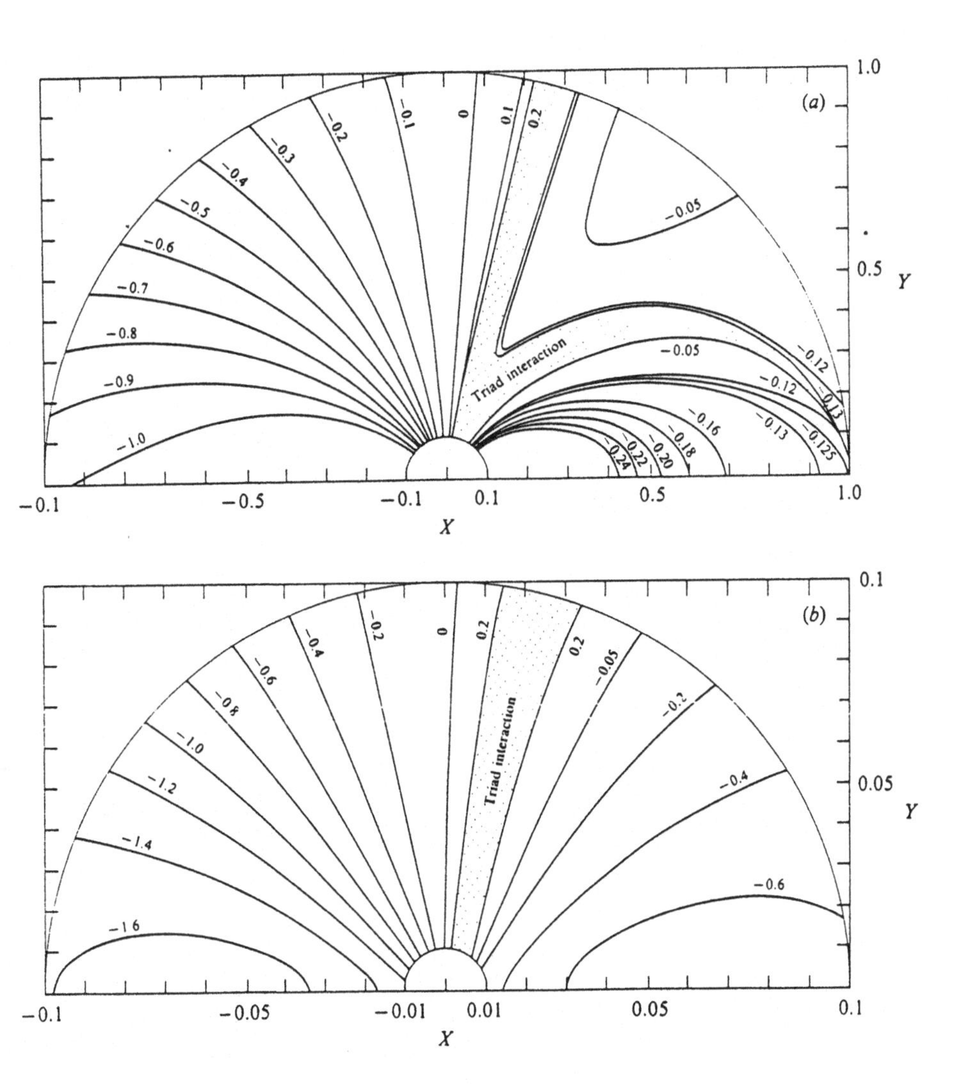

Figure 6 As in Fig. 3 for λ_1 = 0.1cm

Session 4

Perturbation of the gravity-capillary wave spectrum by current variation

Chairperson: W. Alpers

SAR IMAGING OF OCEAN WAVES IN THE PRESENCE OF VARIABLE CURRENTS

Gaspar R. Valenzuela
Space Sensing Branch
Naval Research Laboratory
Washington, DC 20375-5000
USA

ABSTRACT. A review of the hydrodynamic and electromagnetic processes contributing to radar imaging of ocean waves in the presence of variable currents and nonlinear forcing (input from the wind, energy transfer by resonant nonlinear interactions, and dissipation) is given. Since many of the pertinent processes are quite complex and are areas of active research even today, the review is made as comprehensive as possible, but brief enough so as to bring the reader quickly up to date in the literature and still provide enough insight into the processes involved. Some new results on the development of the gravity-capillary wavenumber spectrum are also given.

1. INTRODUCTION

Radar imagery of the ocean (synthetic-SAR and real-RAR) contains a variety of ocean surface features (it is well known that microwaves decay rapidly when propagating into sea water with a skin depth in the order of centimeters). The spatial size of the ocean features imaged ranges from the mesoscale (~100 km) to gravity waves (~100 m), these latter depending on the radar system and wave field parameters. Examples of Seasat SAR imagery of the ocean are given in Beal et al. (1981) and Fu and Holt (1982). See Figure 1.

The radar imagery intensity (contrasts) is the result of electromagnetic (e.m.) fields backscattered by the dynamical ocean surface (a time dependent boundary value problem) after appropriate signal processing and recording (see Rotheram, 1983; Hasselmann et al., 1985). The ocean surface wave field is being forced by wind and other external forces, and is modified during propagation by refraction, nonlinear interactions, dissipation, turbulence, straining by currents, and the influence of other environmental parameters (Hasselmann, 1968).

A number of ocean surface signatures in radar imagery of shallow water (generally less than 40 m deep) relate to bottom topography. G. P. de Loor and H. W. Brunsveld van Hulten (1978) were the first to report features in airborne RAR relating to bottom topography in the

G. J. Komen and W. A. Oost (eds.), Radar Scattering from Modulated Wind Waves, 141–153.

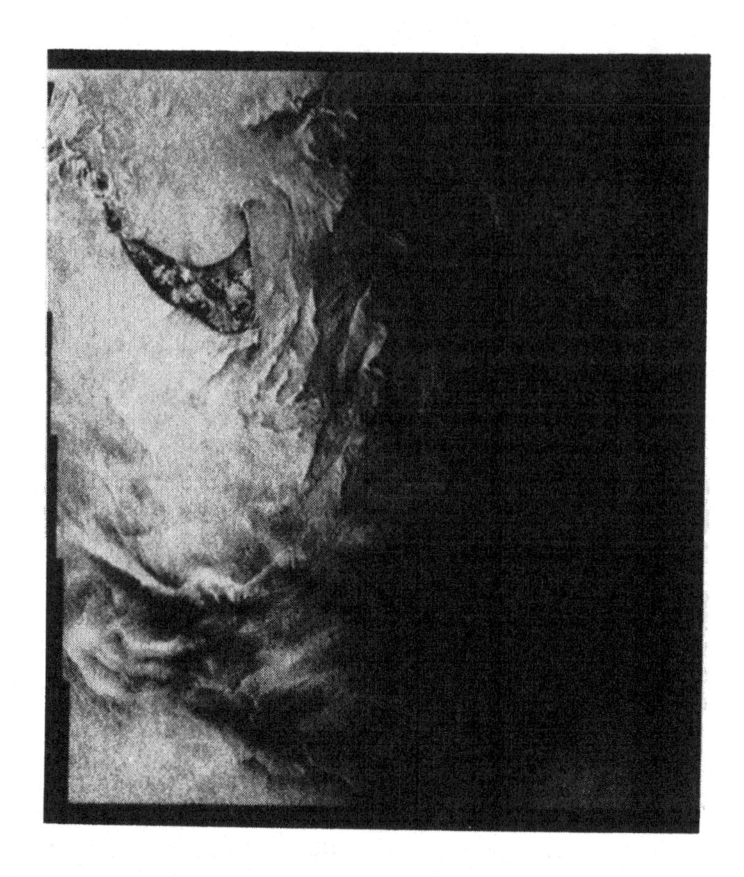

Figure 1. Seasat SAR imagery (L band) of Nantucket Shoals, evident are surface expressions relating the bottom topography, internal waves, current rips, and a front. All these features are the result of current-wave interaction modulating the energy of the Bragg resonant waves of length 30 to 40 cm. Also, notice the dark region (of low backscatter) due to cold upwelling.

North Sea. At first, this fact was a real puzzle since microwaves can not probe the bottom directly at that depth. However, recently the physics involved in the generation of surface signatures relating the bottom topography has become more lucid (Valenzuela et al., 1983; Alpers and Hennings, 1984; Phillips, 1984; Valenzuela et al., 1985).

In coastal areas where these features have been observed with radar imagery, elliptical semi-diurnal M2 tides of 1 m/s (or so) dominate and are modified spatially by the local bottom topography (Zimmerman, 1980; Robinson, 1981; also see Greenberg et al. in this book). Hence, the resulting nonuniform tidal flow is directly related to two-dimensional depth changes, and in turn, the resulting surface current refract the propagation and modulate the energy of the surface wave field. In particular, the energy of short gravity waves of wavenumber $2k_o\sin\theta$ travelling along the line-of-sight of the radar (the Bragg resonant condition, where k_o is the wavenumber of the e.m. radiation in free space and θ is the local angle of incidence of the e.m. wave propagation vector with the local normal vector to the mean surface displacement of the Bragg waves). The Bragg resonant waves are the main contributors of e.m. backscatter power away from nadir and not too close to grazing incidence (Valenzuela, 1978a). Of course, other mechanisms also contribute to the backscatter fields/power (specular, wedge, wave breaking, etc.) in particular towards normal incidence, and for microwaves above 6-10 GHz (Valenzuela, 1985; Holliday et al., 1986). Recall that specular scattering is proportional to the probability density function of slopes evaluated at the specular points, these are surface facets perpendicular to the incident e.m. wave propagation vector with large radii of curvature compared to the radar wavelength.

The fact is that a large number of ocean surface features in radar imagery are the result of current-wave interaction. In this paper, we review the main processes contributing in radar imaging of the ocean surface in the presence of nonuniform currents, and present some preliminary results for the development of the two-dimensional wavenumber spectrum of gravity-capillary waves with time.

2. ELECTROMAGNETIC SCATTERING

Electromagnetic scattering theories for deterministic and statistically rough surfaces are well developed, see reviews by Valenzuela (1978a), Bass and Fuks (1979), and DeSanto and Brown (1986), and of course, numerical results are always possible if the surface is known. However, for geophysical surfaces, as in the case of the nonlinear, time dependent ocean surface, only the first few moments of the statistics of the surface displacement are known, and approximations for the e.m. fields on the surface are necessary. These are the high frequency methods (Physical Optics/Kirchhoff) and the low frequency method (perturbation/Bragg scattering), or combinations of these methods (the two-scale model) (Valenzuela, 1978b).

Recent application of the Kirchhoff approximation (Holliday et al., 1986, 1987) and the two-scale model to SARSEX internal wave imagery data (Thompson, 1988; Thompson et al., 1988) have shown that both of these scattering models, if properly applied, can explain L and X band imagery intensity of internal waves within a couple of decibels. Clearly, this is one more illustration of the accuracy achievable in available scattering models for the ocean. In the past, many tests have been made of the scattering models to other radar frequencies and other oceanographic features, and in the future, many additional quantitative tests will be made to fully assess the capability of available scattering models. Fortunately, the more rigorous Integral Equations and Green's Function method are also available for more accurate predictions of the e.m. scattering from the ocean if these can be validated with improved measurements. For example, spatial and time dependent Green's Function methods including Feynman diagrammatic techniques to sum infinite perturbation series have been developed for wave propagation in random media, and now these are being extended to the ocean surface (DeSanto, 1979; Rotheram, 1983), and to land and sea-ice (Lee and Kong, 1985a, 1985b, 1985c).

2.1. Radar Imaging Theory

The principles of SAR imagery are well established (Harger, 1970; Tomiyasu, 1978). In contrast to RAR (only the amplitude of the return is used) where the azimuthal resolution depends on the antenna beamwidth, the fine azimuthal resolution of SAR is achieved by matched filtering the return signal (amplitude and phase) using the transfer function of SAR. In effect, the known quadratic phase errors across the antenna beam are removed to synthesize a fine azimuthal resolution on the ground/ocean of the order of the azimuthal antenna dimension. As usual, the resolution in range for both RAR and SAR depends on the radar pulse width.

However, for the dynamical ocean surface, additional unknown phase errors are introduced by wave motions and currents which are not removed by matched filtering, and these phase errors cause distortion in the radar imagery intensity. The relationship between the SAR image intensity contrasts and the dynamical ocean surface imaged depends on the nature of the spatial and temporal properties of the backscattered fields from the dynamical ocean surface (a time dependent boundary value problem), see Hasselmann et al. (1985). For the simple case of RAR, the imagery intensity is a direct mapping of the spatial cross-section of the ocean surface averaged over the resolution cell.

Two basic types of SAR models have been developed to predict SAR imagery of the ocean: "velocity bunching" (VBM) and "corrugated" (CM). VBM (or Bragg scattering) models were introduced by Larson et al. (1976), and have been studied in detail by a number of workers. See the review of Alpers et al. (1981), and Plant and Keller (1983). In the VBM, Bragg scattering with spatial decorrelation on the ocean surface is assumed. The Bragg resonant waves travel at their intrinsic phase speed (free waves) and are advected in phase and modulated in energy by straining by the orbital velocity of the dominant waves on the ocean. Local radial velocities cause azimuthal displacements of the surface in the imagery and local accelerations contribute to degradation in spatial resolution. The VBM predicts optimum contrast in the imagery intensity when the SAR is processed with a focus shift (in relation to the optimum focus for a stationary surface) proportional to the maximum radial acceleration on the surface. Monte Carlo simulations of SAR imagery with the VBM have been made by Alpers (1983) for azimuthal waves and by Alpers et al.

(1986) for the two-dimensional spectrum. These latter simulations compare well with spectra obtained with actual SAR imagery.

On the other hand, the CM for SAR imagery also uses Bragg scattering, but assumes the scatterers on the surface are spatially correlated (Harger,1980; Jain, 1981). In the CM, the best contrast in the SAR imagery intensity is achieved for focus shifts from the stationary case which are proportional to the azimuthal component of the phase speed of the dominant waves on the ocean. Validation of the VBM and CM is presently being performed in comparison with data from SAR measurements.

Other more general SAR models have also been developed that assume partial, temporal, and spatial correlation of scatterers on the ocean and these predict maximum contrasts in the SAR imagery intensity when the focus shift is related to the azimuthal component of the phase speed (for the entire surface) and the group speed of the dominant waves when considering the imagery intensity contrast for one facet of the dominant ocean waves (Ivanov, 1982; Rotheram, 1983). However, accurate prediction of SAR imagery of the ocean does not only require a rigorous e.m. scattering/imaging model, but also requires an accurate and realistic representation of the dynamical ocean surface.

3. SURFACE WAVE DYNAMICS

The evolution of the two dimensional spectrum $F(\mathbf{k},\mathbf{x},t)$ of the wave field may be obtained with the conservation of wave action N equation (Hasselmann et al., 1973). In tensor notation, we have

$$\frac{DN}{Dt} = \left(\frac{\partial}{\partial t} + \frac{dx_i}{dt}\frac{\partial}{\partial x_i} + \frac{dk_i}{dt}\frac{\partial}{\partial k_i} \right) N\ (\mathbf{k},\mathbf{x},t) = \frac{S}{\omega} \quad , \tag{3.1}$$

where $N = \omega F/k = E/\omega$ (E is the energy density), $\mathbf{k}$ is the wavenumber vector, $\mathbf{x}$ is the spatial coordinate, t is time, S is the source function, ω is the intrinsic frequency, so that $\Omega = \omega + \mathbf{k}\cdot\mathbf{U}$ is the frequency in a stationary reference frame and $\mathbf{U}$ is the local current. As usual, the wave field propagation is along trajectories determined from the equations

$$\left. \begin{aligned} \frac{dx_i}{dt} &= \frac{\partial\Omega}{\partial k_i} = \frac{\partial\omega}{\partial k_i} + U_i \\ \frac{dk_i}{dt} &= - \frac{\partial\Omega}{\partial x_i} \end{aligned} \right\} \quad . \tag{3.2}$$

In terms of the wavenumber spectrum, (3.1) becomes

$$\left[\frac{\partial}{\partial t} + \frac{\partial\Omega}{\partial k_i}\frac{\partial}{\partial x_i} - \frac{\partial\Omega}{\partial x_i}\left(\frac{\partial}{\partial k_i} + \frac{1}{\omega}\frac{\partial\omega}{\partial k_i} - \frac{1}{k}\frac{\partial k}{\partial k_i} \right) \right] F = \frac{kS}{\omega^2} \quad , \tag{3.3}$$

with the source function to be specified.

The longer gravity waves with dispersion $\omega^2 = gk \tanh(kH)$ (where g is gravity and $H(\mathbf{x})$ is the local depth) are refracted and modulated in amplitude (energy) by variations in depth (Young, 1988) and by current variations (Sakai et al., 1983). On the other hand, the shorter gravity-capillary waves (the Bragg resonant waves) with dispersion $\omega^2 = gk + Tk^3$ (T is surface tension over water density) do not feel the bottom, but are refracted and modulated in amplitude (energy) by the orbital velocity of longer ocean waves (Keller and Wright, 1975; Alpers and Hasselmann, 1978; Phillips, 1981; Sheres, in this book) and by variable currents (Hughes, 1978; Alpers and Hennings, 1984; Phillips, 1984). These previous works only deal with horizontal surface strains, while nonuniform currents may also have vertical shears in many instances (for example, when wind drift is present). Then the current-wave interaction problem is a great deal more complex (Hasselmann, 1968).

Nevertheless, for surface strains, the simplest solution of (3.1) or (3.3) is by perturbation and with the relaxation approximation for the source function (Keller and Wright, 1975; Alpers and Hasselmann, 1978; Alpers and Hennings, 1984). In the relaxation approximation, it is assumed the response of the wave system to straining is a simple exponential relaxation back to equilibrium.

If $F(\mathbf{k},\mathbf{x},t)$ and $\mathbf{U}(\mathbf{x},t)$ are expanded in perturbation series in terms of the fluctuations of $\mathbf{U}$, and $F^{(i)}$, $\mathbf{U}^{(i)}$ are the ith fluctuations of F and $\mathbf{U}$ respectively, then in the relaxation approximation, the source function on the r.h.s. of (3.3) takes the form

$$S = -\beta_r F^{(1)} , \qquad (3.4)$$

with β_r being the relaxation constant. Substituting the perturbation expansions for F and $\mathbf{U}$ in (3.1) or (3.3), and for steady state, the first order solution for the fluctuating part of the spectrum is given by (Alpers and Hennings, 1984)

$$F^{(1)}(\mathbf{k}_s) = \left\{ \frac{\mathbf{k}_L\cdot(\mathbf{c}_{gs} + \overline{\mathbf{U}}) + i\beta_r}{\beta_r^2 + [\mathbf{k}_L\cdot(\mathbf{c}_{gs} + \overline{\mathbf{U}})]^2} \right\} \left(\frac{\partial \mathbf{U}^{(1)}}{\partial x} \cdot \mathbf{k}_s \right) \left(\hat{\mathbf{k}}_L \cdot \hat{\mathbf{k}}_s \right) \left(\frac{\partial F^{(0)}}{\partial k_s} - \eta \frac{F^{(0)}}{k_s} \right) , \qquad (3.5)$$

where $F^{(o)}(\mathbf{k}_s)$ is the unperturbed spectrum for $\mathbf{U} = \mathbf{U}^{(o)} = \overline{\mathbf{U}}$, $k_L = 2\pi/L$ is the wavenumber relating to the spatial scale L of the fluctuations of $\mathbf{U}^{(1)}$, $\mathbf{k}_s$ is the wavenumber of the gravity-capillary waves, $\hat{\mathbf{k}}_p$ is the unit vector of $\mathbf{k}_p$, $\mathbf{c}_{gs}$ is the group speed of the gravity-capillary waves and η is the straining constant. For the straining of a gravity-capillary wave by the orbital velocity of a long gravity wave again $F^{(1)}$ is given by (3.5), except that $\overline{\mathbf{U}}$ is replaced by $-\mathbf{C}_L$ (the phase speed of the gravity wave).

A more exact solution of (3.1) or (3.3) has shown that $\beta_r = \beta_s$ (the wind growth rate of the gravity-capillary waves) (Hughes, 1978), and in the finite amplitude eigenvalue problem of the interaction of long and short water waves including nonlinear resonant interactions it has been shown that $\beta_r \leq \beta_s$ and to be dependent on the magnitudes of the spectra of the wave fields (Valenzuela and Wright, 1979). Bagg et al. (1986) used the concept of passage for the current-wave interaction problem and generalized Hughes (1978) solution to arbitrary power for the spectral dissipation and demonstrated that the adiabatic case (S = 0) yields the largest modulation of the wave spectrum $F(\underset{\sim}{k},\underset{\sim}{x},t)$ (see Thomas et al. in this book).

Hence, first order variations in radar cross-section of the ocean in the presence of currents may be obtained from (3.5) using Bragg scattering, and SAR imagery intensity variations are related to (3.5) if hydrodynamic effects dominate, otherwise tilts and velocity bunching contributions also have to be included. Using conservation of mass for water flowing over a bottom ridge or bank, the current gradient normal to the ridge was obtained by Alpers and Hennings (1984), and given by the expression

$$\frac{\partial \underset{\sim}{U}^{(1)}}{\partial x} \propto \frac{\overline{\underset{\sim}{U}}}{H^2} \frac{\partial H}{\partial x_\perp} , \tag{3.6}$$

where $x_\perp$ is normal to the bottom ridge or bank. Therefore, to first order the radar imagery intensity is related to changes in depth and bottom slopes. Of course, in the two-dimensional problem, not only current gradients transverse to the bottom ridge contribute to the hydrodynamic modulations of Bragg resonant wave energy, but in addition, longitudinal current gradients contribute as well. Numerical results for the two-dimensional tidal flow in Nantucket Shoals have been obtained by Greenberg et al. (in this book).

3.1. Energy Balance of Short Gravity Waves

The energy balance of gravity waves has been investigated in detail by a number of workers, in particular see Hasselmann et al. (1973, 1976) and the SWAMP Group (1985) for tests on several numerical models. Komen et al. (1984) obtained the conditions for the existence of the fully developed two-dimensional wind-sea spectrum, and the energy balance of gravity waves for finite-depth was studied by Weber (1988). Models for shallow water have also been developed and tested, see for example the SWIM Group (1985) and Young (1988).

In relation to the short gravity waves, Plant (1980) investigated in a wavetank their energy balance and found that, in addition to wind input and dissipation, both nonlinear energy transfers by resonant interactions (quartets) (Hasselmann, 1963) and (triads) (Valenzuela and Laing, 1972) contributed. However, for the gravity-capillary resonant interactions, the wind drift had to be included in the dispersion relation; the order of magnitude of quartets and triads resonant interactions was about equal for 10 cm length waves. Also see Valenzuela and Wright (1979).

Recently, van Gastel (1987a, 1987b) has investigated the development of the gravity-capillary wave spectrum (in one dimension) for forcing by wind, energy transfer by triad resonant interactions, and dissipation (wave-breaking and viscosity) and obtained a multi-peaked spectrum. Here, we have obtained preliminary results for the development of the two-dimensional gravity-capillary wave spectrum with time. In our preliminary results, we assume separability (i.e., $F(\mathbf{k}) = F(k,\phi) = F(k)F(\phi)$ and have let $F(\phi) = 2/\pi \cos^2\phi$ for the angular spreading function). For the source function, $S = S_{in} + S_{vi} + S_{nl} + S_{wb}$ we have selected expressions similar to van Gastel's (1987b), but have adjusted their strength to match Plant's (1980) measurements for the magnitude of the equilibrium spectrum. Accordingly, we have used the following expressions:

$$\text{(wind input)} \quad S_{in} = 0.04\left(\frac{u_*}{c_s}\right)^2 \omega \cos\phi \, F(\mathbf{k}) \tag{3.7}$$

$$\text{(viscous)} \quad S_{vi} = -4\nu k^2 \, F(\mathbf{k}) \tag{3.8}$$

$$\text{(nonlinear)} \quad S_{nl} = 2 \text{ x (eq. 4.4 of Valenzuela \& Laing, 1972)} \tag{3.9}$$

$$\text{(wave breaking)} \quad S_{wb} = -0.005\left(\frac{\omega}{\omega_p}\right)^q \left(\frac{\alpha}{0.01}\right)^2 F(\mathbf{k}) \tag{3.10}$$

where q is as in van Gastel (1987b) and

$$\alpha = \left[\frac{\int k^2 F(\mathbf{k})\, d\mathbf{k}}{2\pi^2}\right]^{1/2} .$$

Other parameters are as follows: u_* is the air friction velocity, c_s is the phase speed of gravity-capillary waves, ν is the kinematic viscosity of water and ω_p is the frequency of the peak of the spectrum.

The numerical results were obtained with a Cray X-MP/24 computer using time increments of 0.02 s, and 200 increments in wavenumber in the interval 0.1 k_m and 10 k_m (k_m is the wavenumber of 1.7 cm waves), and a k^{-4} fall-off rate for the spectrum was used for $k > 10\ k_m$. The development of the spectrum with time for $u_* = 20$ cm/s is shown in Figure 2 and the corresponding source function in Figure 3. Equilibrium was approached after 2 sec., oddly enough the peak of the spectrum shifts towards shorter waves rather than to longer waves. This could be a problem of resolution in wavenumber in the calculations, however, contrary to van Gastel's (1987b) the spectra are fairly smooth in these two-dimensional calculations.

Presently, we are in the process of implementing the numerical code for the two-dimensional development of the gravity-capillary wave spectrum in a parallel processor (the Connection Machine) where we should have a great deal more flexibility in exploring in more detail this problem, and where ultimately we should include in the source function, both the nonlinear energy transfer by triads (Valenzuela and Laing, 1972) and quartets (Hasselmann, 1963), and, of course, currents if possible.

Van Gastel (1987b) has, in addition, investigated the development of gravity-capillary waves in the presence of variable currents and

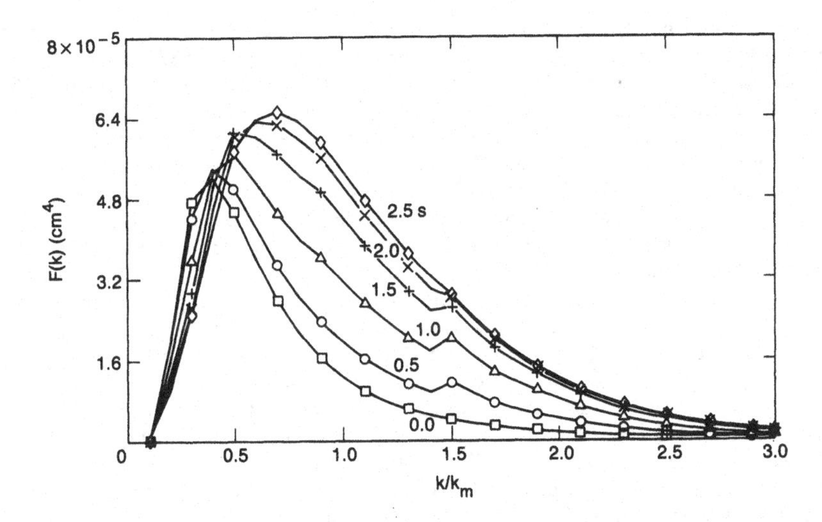

Figure 2. Development with time of the gravity-capillary wavenumber spectrum for forcing by wind, energy transfer by resonant nonlinear interactions, viscous dissipation, and wave-breaking. k_m is the wavenumber of 1.7 cm waves (3.696 cm^{-1}).

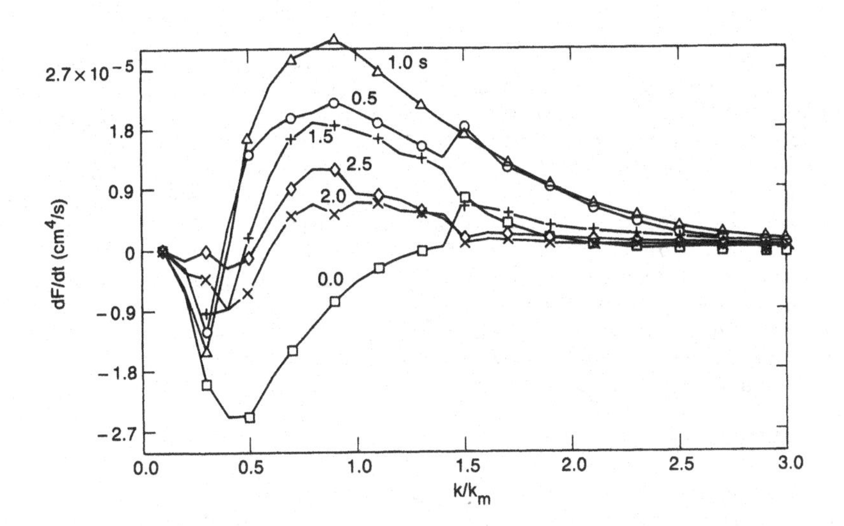

Figure 3. The source function as a function of time used in the development of the gravity-capillary wavenumber spectrum.

has found unusually large (up to 30 db) modulation of the energy of the wavenumber spectrum of gravity-capillary waves for light to moderate winds. The nonlinear forcing used could be one reason for these large modulations, however, Bagg et al. (1986) indicate that the refraction term in (3.1) or (3.3) also is responsible for large modulations in the spectrum. We recall that in Nantucket Shoals (Valenzuela et al., 1985) unusually large variations in radar backscatter were observed for light winds for X band radar measurements. Another mechanism that also can cause unusually large modulations of the wave spectrum is atmospheric stability (Valenzuela and Chen, 1985) where it was shown that stable atmospheric conditions (warm air over cold water) can also cause large modulations in the radar cross-section of the sea. This can be shown with (3.5) if we let $\beta_r = \beta_s >> \underset{\sim}{k}_L \cdot [\underset{\sim}{c}_{gs} + \underset{\sim}{U}]$ and include the effect of atmospheric stability on β_s which is proportional to wind stress ~ u_*^2. However, the maximum modulations achievable are constrained by the adiabatic solution (S = 0) in (3.1) or (3.3), see Bagg et al. (1986) or Thomas et al. in this book. Also see Keller et al. (1985) for measurements on the atmospheric stability dependence of the radar cross-section of the sea.

4. CONCLUSIONS

Although the hydrodynamics and e.m. scattering processes contributing to radar imaging of ocean waves in the presence of variable currents are quite complex, a great deal of knowledge and understanding of the processes has been attained in the recent years as discussed in the review.

However, a great deal of wave research is needed on a number of processes, in particular in the development (with time and fetch) of the two-dimensional wavenumber spectrum of short gravity waves in the presence of longer gravity waves and variable currents, even with vertical shear as in the case of wind drift. Additional information is also needed on the impact of a number of environmental parameters (such as wind, turbulence, dissipation, nonlinear forcing, etc.) on the development of the short wind wave spectrum.

Furthermore, validation of e.m. scattering models is needed to test the accuracy of the two-scale model, Kirchhoff approximation and others in the prediction and interpretation of radar imagery for frequencies above X band. See the recommendations given in the General Discussion of this book.

ACKNOWLEDGMENTS

The author would like to thank Sanjay S. Natarajan of Carnegie-Mellon University who participated in the Naval Research Laboratory Summer Program of 1987 and performed the numerical calculations on the development of the gravity-capillary wave spectrum.

REFERENCES

Alpers, W., 1983: Monte Carlo simulations for studying the relationship between ocean wave and synthetic aperture radar image spectra. J. Geophys. Res., **88**, 1745-1759.

Alpers, W. and K. Hasselmann, 1978: The two-frequency microwave technique for measuring ocean wave spectra from the airplane or satellite. Boundary Layer Meteorol., **13**, 215-230.

Alpers, W. R. and I. Hennings, 1984: A theory of the imaging mechanism of underwater bottom topography by real and synthetic aperture radar. J. Geophys. Res., **89**, 10529-10546.

Alpers, W., D. B. Ross and C. L. Rufenach, 1981: On the detectability of ocean surface waves by real and synthetic aperture radar. J. Geophys. Res., **86**, 6481-6498.

Alpers, W., C. Bruening and K. Richter, 1986: Comparison of simulated and measured synthetic aperture radar image spectra with buoy-derived ocean wave spectra during the Shuttle Imaging Radar B mission. IEEE Trans., **GE-24**, 559-566.

Bagg, M. T., A. C. Edwards, J. R. Perry, J. C. Scott, J. A. Stacey and J. O. Thomas, 1986: The SIR-B Mission: Towards an Understanding of Internal Waves in the Ocean. ARE TR-86122, Admiralty Research Establishment, Portland Dorset, 50 pp.

Bass, F. G. and I. M. Fuks, 1979: Wave Scattering from Statistically Rough Surfaces. Pergamon Press, 525 pp.

Beal, R. C., P. S. DeLeonibus and I. Katz, 1981: Spaceborne Synthetic Aperture Radar for Oceanography. Johns Hopkins Press, 215 pp.

de Loor, G. P. and H. W. Brunsveld van Hulten, 1978: Microwave measurements over the North Sea. Boundary-Layer Meteorol., 13, 119-131.

DeSanto, J. A., 1979: Coherent scattering from rough surfaces. Mathematical Methods and Applications of Scattering Theory., J. A. DeSanto, A. W. Saenz, and W. W. Zachary, Eds., Springer-Verlag, 60-70.

DeSanto, J. A. and G. S. Brown, 1986: Analytical techniques for multiple scattering from rough surfaces., Progress in Optics, 23, E. Wolf, Ed., 1-62.

Fu, L.-L. and B. Holt, 1982: Seasat Views Oceans and Sea Ice with Synthetic Aperture Radar. NASA/JPL Publication 81-120, Jet Propulsion Laboratory, Pasadena, CA 200 pp.

Gastel, K. van, 1987a: Nonlinear interactions of gravity-capillary waves: Lagrangian theory and effects on the spectrum. J. Fluid Mech., **182**, 449-523.

Gastel, K. van, 1987b: Imaging of X band radar of subsurface features: a nonlinear phenomenon. J. Geophys. Res., **92**, 11857-11865.

Harger, R. O., 1970: Synthetic Aperture Radar Systems. Academic, 240 pp.

Harger, R. O., 1980: The synthetic aperture radar image of time-variant scenes. Radio Science, **15**, 749-657.

Hasselmann, K., 1963: On the non-linear energy transfer in a gravity wave spectrum. Part 3. J. Fluid Mech., 15, 385-398.

Hasselmann, K., 1968: Weak-interaction theory of ocean waves. Basic

Developments in Fluid Dynamics, M. Holt, Ed., Academic, 117-182.
Hasselmann, K. et XV al., 1973: Measurements of wind-wave growth and swell decay during the Joint North Sea Wave Project (JONSWAP). Dtch. Hydrogr. Z., **A8**, No. 12, 95 pp.
Hasselmann, K., R. K. Raney, W. J. Plant, W. Alpers, R. A. Schuchman, D. R. Lyzenga, C. L. Rufenach and M. J. Tucker, 1985: Theory of synthetic aperture radar ocean imaging: a MARSEN view. J. Geophys. Res., 90, 4659-4686.
Holliday, D., G. St-Cyr and N. E. Woods, 1986: A radar ocean imaging model for small to moderate incidence angles. Int. J. Remote Sensing, **7**, 1809-1834.
Holliday, D., G. St-Cyr and N. E. Woods, 1987: Comparison of a new radar ocean imaging model with SARSEX internal wave image data. Int. J. Rem. Sens., **8**, 1423-1430.
Hughes, B. A., 1978: The effect of internal waves on surface wind waves 2. theoretical analysis. J. Geophys. Res., **83**, 455-465.
Ivanov, A. V., 1982: On the synthetic aperture radar imaging of ocean surface waves. IEEE Trans., **OE-7**, 96-103.
Jain, A., 1981: SAR imaging of ocean waves: theory. IEEE Trans., **OE-6**, 130-139.
Keller, W. C. and J. W. Wright, 1975: Microwave scattering and the straining of wind-generated waves. Radio Sci., **10**, 139-147.
Keller, W. C., W. J. Plant and D. E. Weissman, 1985: The dependence of X band microwave sea return on atmospheric stability and sea state. J. Geophys. Res., **90**, 1019-1029.
Komen, G. J., S. Hasselmann and K. Hasselmann, 1984: On the existence of a fully developed wind-sea spectrum. J. Phys. Oceanogr., **14**, 1271-1285.
Larson, T. R., L. I. Moskowitz and J. W. Wright, 1976: A note on SAR imagery of the ocean. IEEE Trans., **AP-24**, 393-394.
Lee, J. K. and J. A. Kong, 1985a: Active microwave remote sensing of an anisotropic random medium layer. IEEE Trans., **GE-23**, 910-923.
Lee, J. K. and J. A. Kong, 1985b: Passive microwave remote sensing of an anisotropic random medium layer. IEEE Trans., **GE-23**, 924-932.
Lee, J. K. and J. A. Kong, 1985c: Electromagnetic wave scattering in a two-layer anisotropic random medium. J. Opt. Soc. Am. A., **2**, 2171-2186.
Phillips, O. M., 1981: The structure of short gravity waves on the ocean surface. Spaceborne Synthetic Aperture Radar for Oceanography, R. C. Beal, P. S. DeLeonibus and I. Katz. Eds., Johns Hopkins Press, 24-31.
Phillips, O. M., 1984: On the response of short ocean wave components at a fixed wavenumber to ocean current variations. J. Phys. Oceanogr., 14, 1425-1433.
Plant, W. J., 1980: On the steady-state energy balance of short gravity wave systems. J. Phys. Oceanogr., **10**, 1340-1352.
Plant, W. J. and W. C. Keller, 1983: The two-scale radar wave probe and SAR imagery of the ocean. J. Geophys. Res., **88**, 9776-9784.
Robinson, I. S., 1981: Tidal vorticity and residual circulation. Deep Sea Res., **28A**, 195-212.
Rotheram, S., 1983: Theory of SAR ocean wave imaging. Satellite

Microwave Remote Sensing, T. D. Allan, Ed., Ellis Horwood, 155-186.

SWAMP Group, 1985: Ocean Wave Modeling, Plenum Press, 256 pp.

SWIM Group, 1985: Shallow water intercomparison of wave models. The Ocean Surface, Y. Toba and H. Mitsuyasu, Eds., Riedel, 201-220.

Sakai, T. M. Koseki, and Y. Iwagaki, 1983: Irregular wave refraction due to current. J. Hydr. Eng., **109**, 1203-1215.

Thompson, D. R., 1988: Calculation of radar backscatter modulations from internal waves. J. Geophys. Res., **93**, in press.

Thompson, D. R., B. L. Gotwols and R. E. Sterner, 1988: A comparison of measured surface wave spectral modulations with predictions from a wave-current interaction model. J. Geophys. Res., **93**, in press.

Tomiyasu, K., 1978: Tutorial review of synthetic aperture radar (SAR) with applications to imaging of the ocean surface. IEEE Proc., **66**, 563-583.

Valenzuela, G. R., 1978a: Theories for the interaction of electromagnetic and oceanic waves: review. Boundary-Layer Meteorol., **13**, 61-85.

Valenzuela, G. R., 1978b: Scattering of electromagnetic waves from the ocean. Surveillance of Environmental Pollution and Resources by Electromagnetic Waves, T. Lund, Ed., Riedel, 199-226.

Valenzuela, G. R., 1985: Microwave sensing of the ocean surface. The Ocean Surface, Y. Toba and H. Mitsuyasu, Eds., Riedel, 233-244.

Valenzuela, G. R. and M. B. Laing, 1972: Nonlinear energy transfer in gravity-capillary wave spectra, with applications. J. Fluid Mech., **54**, 507-520.

Valenzuela, G. R. and J. W. Wright, 1979: Modulation of short gravity-capillary waves by longer-scale periodic flows - a higher order theory. Radio Science, **14**, 1099-1110.

Valenzuela, G. R. and D. T. Chen, 1985: The Effect of Atmospheric Stability on the Modulation of Microwave Backscatter. North American Radio Science Mtg. (CNC & USNC/URSI), Vancouver, Canada, 17-21 June.

Valenzuela, G. R., D. T. Chen, W. D. Garrett and J. A. C. Kaiser, 1983: Shallow water bottom topography from radar imagery. Nature, **303**, 687-689.

Valenzuela, G. R., W. J. Plant, D. L. Schuler, D. T. Chen and W. C. Keller, 1985: Microwave probing of shallow water bottom topography in the Nantucket Shoals. J. Geophys. Res., **90**, 4931-4942.

Weber, S. L., 1988: The energy balance of finite depth gravity waves. J. Geophys. Res., **93**, 3601-3607.

Young, I. R., 1988: A shallow water spectral wave model. J. Geophys. Res., **93**, 5113-5129.

Zimmerman, J. T. F., 1980: Vorticity transfer by tidal currents over an irregular topography. J. Marine Res., **38**, 601-630.

THE MODULATION TRANSFER FUNCTION: CONCEPT AND APPLICATIONS

William J. Plant
U.S. Naval Research Laboratory
Washington, D.C. 20375-5000
USA

ABSTRACT. This paper outlines the concept of the modulation transfer function used to relate fluctuations in power received by an active microwave system viewing the ocean to the dominant surface waves on the ocean which cause the fluctuations. Expressions are given for the purely geometric modulations related to surface tilting and changes in range. The manner in which the modulation transfer function enters into the interpretation of real and synthetic aperture radar images, wave spectrometer outputs, scatterometer wind retrievals, and altimeter measurements is reviewed. Finally, limitations of the concept are discussed.

1. INTRODUCTION

The concept of an ocean wave/radar modulation transfer function (or MTF) was first introduced by Keller and Wright (1975) (as the power modulation index) to account for the modulation of microwave backscatter from a small area of rough water by waves much longer than the dimensions of that area. Their idea was that small perturbations in wind wave equilibrium caused by the longer waves could be considered to be of first order in long wave slope. Since its introduction, the MTF has found application in a wide variety of microwave techniques for the study of the air/sea interface. Occasionally, however, some misunderstanding of its definition, measurement, or application is evidenced in the literature. This paper is an attempt to remedy the situation by reviewing the concept, measurement, applications, and limitations of the modulation transfer function.

2. THE CONCEPT OF A MODULATION TRANSFER FUNCTION

Consider the backscattered signal to a microwave radar which illuminates an area of sea surface whose dimension in the direction of propagation of the dominant surface wave is much smaller than the wavelength of that wave. The backscattered power is found to fluctuate

G. J. Komen and W. A. Oost (eds.), Radar Scattering from Modulated Wind Waves, 155–172.

with time and may be represented by a Fourier series:

$$P(t) = \sum_{n=0}^{\infty} P_n \cos(2\pi nft + \Phi_n) \quad . \tag{1}$$

If only a single long wave is present on the sea surface, the low frequency fluctuations of P(t) are primarily associated with the frequency, f_o, of this long wave and we may write,

$$P(t) = P_o \cos\Phi_o + P_1 \cos(2\pi f_o t + \Phi_1) + P_2 \cos(4\pi f_o t + \Phi_2) + \ldots \quad . \tag{2}$$

In most cases, harmonics of the ocean wave frequency are not found in the return power. Furthermore, it is found that the amplitude of the fluctuations in the power increase with the height of the ocean wave. To incorporate these observations in P(t), we let

$$\begin{aligned} P_o \cos\Phi_o &= \bar{P} \\ P_1 &= \bar{P}|R|a \\ \Phi_1 &= \Phi \\ P_n &= 0 \quad , \quad n > 1 \end{aligned} \tag{3}$$

where $\bar{P}$ is mean backscattered power, a is wave amplitude, and R is a coefficient describing the power modulations. With these definitions, we may write

$$P(t) = \bar{P}\,[1 + |R|a \cos(2\pi f_o t + \Phi)] \quad . \tag{4}$$

It is often desirable to relate the fluctuations in scattered power to long wave slope rather than amplitude. If we describe the surface displacement $\gamma(\vec{x},t)$ by the first order expression

$$\gamma(\vec{x},t) = a \cos(\vec{K}\cdot\vec{x} - 2\pi f_o t) \tag{5}$$

then, wave slope is given by

$$\nabla\gamma(\vec{x},t) = \vec{K}a \sin(\vec{K}\cdot\vec{x} - 2\pi f_o t) \quad . \tag{6}$$

In deep water, we may therefore write the magnitude of the wave slope as

$$|\nabla\gamma| = Ka = KCa/C = 2\pi f_o a/C = \frac{U_o}{C} \tag{7}$$

where C is $2\pi f_o/K$, the phase speed of the ocean wave, and

$$U_o = \left| \frac{\partial\gamma}{\partial t} \right| \tag{8}$$

is the magnitude of the horizontal component of orbital velocity. Thus,

$$P(t) = \bar{P}\ [1 + |m|\ (U_o/C)\ \cos(2\pi f_o t + \Phi)] \tag{9}$$

where

$$|m| = |R|/K \tag{10}$$

If we write this in complex notation, we have

$$P(t) = \bar{P}\ [1 + m\ (U_o/C)\ e^{i2\pi f_o t}\ ] \tag{11}$$

where m is now a complex quantity whose phase is positive if it leads the wave crest and zero at the crest. Since m will be a function of ocean wave frequency or direction (as well as other parameters), it is called the modulation transfer function. It is a dimensionless quantity and is related to R, which has dimensions of inverse length, by

$$m = R/K \quad . \tag{12}$$

The function m has been used in publications by researchers from the U.S. Naval Research Laboratory while R is the transfer function used by investigators at the Max-Plank-Institut fur Meteorologie in Germany.

The MTF may be evaluated by cross-correlating P(t) with either wave height or horizontal component of orbital velocity. Correlating P(t) and $\gamma(0,t)$, we have, using complex notation,

$$P(t)*\gamma^*(0,t) = \frac{1}{2}\ \bar{P}Ra^2 e^{-i2\pi f_o \tau} \tag{13}$$

where τ is the time lag. But

$$\gamma(0,t)*\gamma^*(0,t) = \frac{1}{2}\ a^2 e^{-i2\pi f_o \tau} \tag{14}$$

so

$$R = \frac{P(t)*\gamma^*(0,t)}{\bar{P}[\gamma(0,t)*\gamma^*(0,t)]} \quad . \tag{15}$$

The function R may therefore be determined by correlating P(t) with the output of a wave staff or its equivalent.

If instead, we correlate P(t) with the horizontal component of orbital velocity,

$$U(\vec{x},t) = U_o e^{i(\vec{K}\cdot\vec{x} - 2\pi f_o t)} \quad (16)$$

we obtain the function m. This is shown as follows:

$$P(t)*U^*(0,t) = \frac{1}{2} \bar{P}m \, (U_o/C) \, U_o e^{-i2\pi f_o \tau} \quad (17)$$

or,

$$m = \frac{C \, [P(t)*U^*(0,t)]}{\bar{P} \, [U(0,t)*U^*(0,t)]} \quad . \quad (18)$$

This method is convenient when coherent, linearly-detected radars are being used since U(0,t) may be obtained directly from the mean Doppler shift of the backscattered signal. The relation for water of depth d is,

$$U(t) = \left\{ \frac{\lambda_o f_d}{2\sin\theta_o [\sin^2\theta_o \cos^2\phi_o + \cos^2\theta_o \tanh^2 kd]^{1/2}} \right\}$$

$$\exp \left\{ -i \left[2\pi f_o t + \tan^{-1} \left(\frac{\tanh Kd \cot\theta_o}{\cos\phi} \right) \right] \right\} \quad (19)$$

where ϕ_o is the angle between the direction from which the long wave is propagating and the horizontal antenna-look direction, λ_o is the radar wavelength, θ_o is mean incidence angle, and f_d is the mean Doppler shift.

3. THE MODULATION TRANSFER FUNCTION FOR A NATURAL ENVIRONMENT

So far we have only discussed the case where a single long wave exists on the surface. In an ocean environment, of course, this is seldom the case. A more realistic treatment may be developed assuming that long waves cover some range of frequencies and directions. If both dimensions of the illuminated area are small compared to ocean wavelengths, no directional resolution of these waves is possible and the temporally varying backscattered power may be expressed,

$$P(\vec{x}_o,t) = \bar{P} + \int \tilde{P}(K',f,\phi) e^{i(K' x_o \cos\phi - 2\pi f t)} K' dK' df d\phi \quad (20)$$

where $\vec{x}_o$ is the horizontal projection of the range to the center of the illuminated spot and ϕ is the angle between $\vec{K}'$ and this projection. If a dispersion relation exists then,

$$\tilde{P}(K',f,\phi) = \tilde{P}(f,\phi)\delta(K' - K(f))/K' \tag{21}$$

and

$$P(\vec{x}_o,t) = \bar{P} + \int \tilde{P}(f,\phi)e^{i(Kx_o\cos\phi - 2\pi ft)}dfd\phi \quad . \tag{22}$$

Once again, we may relate spectral densities to either wave height or orbital velocity,

$$\tilde{P}(f,\phi) = \bar{P}R(f,\phi)a(f,\phi) = \bar{P}m(f,\phi)U_o(f,\phi)/C(f) \tag{23}$$

where now in the space-time domain,

$$\gamma(\vec{x}_o,t) = \int a(f,\phi)e^{i(Kx_o\cos\phi - 2\pi ft)}dfd\phi \tag{24}$$

and

$$U(\vec{x}_o,t) = \int U_o(f,\phi)e^{i(Kx_o\cos\phi - 2\pi ft)}dfd\phi \quad . \tag{25}$$

Now let $\vec{x}_o = 0$ (this essentially invokes large-scale homogeneity) and define

$$\tilde{P}(f) = \int \tilde{P}(f,\phi)d\phi \tag{26}$$

with similar equations for a(f) and $U_o(f)$. The cross spectrum of P(0,t) and $\gamma(0,t)$ is

$$\tilde{P}(f)a^*(f) = \bar{P}a^*(f) \int R(f,\phi)a(f,\phi)d\phi \tag{27}$$

while that of P(0,t) and U(0,t) is

$$\tilde{P}(f)U_o^*(f) = \bar{P}[U_o^*(f)/C(f)] \int m(f,\phi)U_o(f,\phi)d\phi \quad . \tag{28}$$

In the natural environment, the MTF refers to a weighted directional average. In this case, let $\bar{R}$ and $\bar{m}$ be the two functions defined as the MTF. Then,

$$\bar{R}(f) = \frac{\int R(f,\phi)a(f,\phi)d\phi}{a(f)} = \frac{\tilde{P}(f)a^*(f)}{\bar{P}|a(f)|^2} \tag{29}$$

$$\bar{m}(f) = \frac{\int m(f,\phi)U_o(f,\phi)d\phi}{U_o(f)} = \frac{C(f)[P(f)U_o^*(f)]}{\bar{P}|U_o(f)|^2} \quad . \tag{30}$$

These are the functions usually presented when modulation transfer functions are calculated for the natural environment (Plant, et al., 1978; Wright, et al., 1980; Plant, et al., 1983; Schroter, et al., 1986; Feindt, et al., 1986). If $a(f,\phi)$ does not maximize at $\phi = 0$, then $\bar{R}(f)$ and $\bar{m}(f)$ may also be parametrically dependent on the angle which the dominant wave makes with the horizontal projection of the antenna-look direction. Note that the relationship between $\bar{m}$ and $\bar{R}$ is the same as Eq. 12, namely,

$$\bar{m}(f) = \bar{R}(f)/K(f) \quad . \tag{31}$$

If only the range dimension of the illuminated area were small compared to long wavelengths, then directionality would be achieved since only waves travelling very nearly in the range direction would modulate the backscattered power (the footprint averages over crests and troughs of other waves yielding little time variation). In such a case,

$$P(\vec{x}_o,t) = \bar{P} + \int P(f,0)e^{i(\vec{K}\cdot\vec{x}_o - 2\pi ft)}df \tag{32}$$

and

$$U(\bar{x}_o,t) = \int U_o(f,0)e^{i(\vec{K}\cdot\vec{x}_o - 2\pi ft)}df \tag{33}$$

if U is derived from the FM part of the radar return. Then only $\bar{m}(f,0)$ will be obtained from cross-spectral analysis:

$$\bar{m}(f,0) = \frac{C(f)[\tilde{P}(f,0)U_o^*(f,0)]}{\bar{P}|U(f,0)|^2} \tag{34}$$

Data cannot usually be collected in this manner from towers and piers where antennas are close to the sea surface.

4. PRACTICAL CONSIDERATIONS OF MEASUREMENTS AND THEORY

The last section showed that R could be computed from cross-spectra of backscattered power with a signal related to waveheight. Similarly m could be computed from cross-spectra of power with a signal related to the horizontal component of orbital velocity. An important quantity in analyses of this type is the coherence function defined by

$$\gamma^2(f) = \frac{|\tilde{P}(f)g^*(f)|^2}{|\tilde{P}(f)|^2\ |g(f)|^2} . \tag{35}$$

In MTF calculations, g(f) is either a(f) or $U_0(f)$. The coherence function has a maximum value of one at any given frequency. Values less than one indicate either (a) noisy signals, (b) P(f) and g(f) are not linearly related, or (c) P(f) is due to sources in addition to g(f). In backscattering measurements in a natural environment, $\gamma^2(f)$ is typically less than one, sometimes much less. A useful, though arbitrary convention in analyzing these data is to use only data for which $\gamma^2(f) > 0.3$.

One result of low values of $\gamma^2(f)$ is that corresponding values of m or R have large variances. For instance, if $\gamma^2(f) = 0.3$ and 20 spectra are averaged in determining m or R, then 95% confidence intervals on the magnitudes of the functions are nearly 90% of the mean values. Thus many more than 20 spectra must be averaged in order to obtain accurate values of m or R. In most cases, however, m and R are functions of environmental parameters such as wind speed and, furthermore, records of about one minute in length are necessary to obtain adequate spectral resolution. Thus 20 spectra require about 20 minutes (less if overlapping is allowed) data records. This is about the maximum time over which environmental parameters can be assumed stationary. The common procedure, therefore, is to calculate values of m or R from about 20 spectra obtained from about 20 minutes of data and then average many such values derived from data collected under similar environmental conditions. This procedure allows one to obtain accurate values of m(f) for various environmental conditions.

If the modulation of the scattered power is not linearly related to waveheight or slope, then derived values of m or R should be functions of waveheight or slope. In fact, such dependence has been found at intermediate incidence angles for microwave frequencies higher than L-Band (1.5 GHz) along with the standard dependence of the MTF on wind speed (Keller, et al. 1985). The effect in either case is a decrease of the MTF with an increase of the independent variable, either wind speed or wave slope. Since the definition of the MTF requires division by the average received power, the observed dependence of the MTF on wind speed and wave slope is consistent with the observation that the cross section, which is proportional to the average received power, increases with increases in these two variables. Thus a consistent picture of microwave backscatter from the sea surface at intermediate incidence angles seems to be that the average backscattered power increases with wind but that the peak-to-peak modulation of the received power causes by long waves remains nearly constant with wind speed; the ratio of peak-to-peak modulation to average received power level then decreases with wind speed. Similarly, present indications are that cross sections increase with long wave slope due to second-order effects of long-wave modulation. If only effects up to second order in long-wave slope are important in backscattering, then the peak-to-peak modulation will depend on slope to the first power so the observed MTF decreases with wave slope due to second order effects on the cross section. If these indications are correct, the concept of an MTF linearly relating wave slope or height to received power can be retained but must be allowed to vary with wave slope if normalized by average received power.

The MTF as defined in the above sections is an ocean wave/radar MTF and includes all effects by which long ocean waves cause changes in the level of power received by the microwave system. Working within a Bragg-scattering/composite-surface formulation of microwave backscattering, the different possible effects may be readily defined. These include the actual hydrodynamic modulation of the spectral density of the Bragg wave as well as the purely geometric effects of surface tilt and range change. The latter two effects may be readily understood. Microwave backscatter from the sea surface is stronger for smaller incidence angles. Thus surface tilts caused by the long waves modulate the backscatter by alternately increasing and decreasing the local incidence angle. This effect is always in phase with or 180 degrees out of phase with the slope of the long wave depending whether the wave travels toward the antenna or away from it. Similarly, the change in surface elevation caused by the long wave changes the range to the antenna thus alternately increasing and decreasing the power received by the microwave system. This source of modulation will always be in phase with the wave amplitude.

If one wants to investigate the hydrodynamic modulation of short waves by long then components of the MTF caused by geometric factors must be removed from the measured values. If we write

$$m = m' - it + r \tag{36}$$

where m′ is the hydrodynamic part of the MTF, t is the tilt part, and r is the range-change part, then we may evaluate the latter two factors using composite-surface theory. According to this theory, the received power is given by

$$P \propto \frac{\sigma_o A}{R_i^4} = \frac{(16\pi k_o^4 G\psi)\ A}{R_i^4} \tag{37}$$

where σ_o is normalized radar cross section, k_o is microwave number, $G(\theta)$ is a function of incidence angle dependent on polarization, $\psi(k)$ is short-wave spectral density as a function of short wave number, R_i is range to the surface, and A is illuminated area. For large dielectric constants such as that of the sea surface, where it is about 81, the function G is given approximately for vertical polarization by

$$G(\theta) = \frac{\cos^4\theta(1 + \sin^2\theta)^2}{(\cos\theta + .111)^4} \tag{38}$$

and for horizontal polarization by

$$G(\theta) = \frac{\cos^4\theta}{(.111\cos\theta + 1)^4} \tag{39}$$

An expression for the illuminated area depends on whether the footprint on the surface is defined in the range direction by the antenna beamwidth or by the pulse length of the system. We have

$$A = \begin{cases} \dfrac{1}{2} R_i^2 \Phi_V \Phi_H / \cos\theta , & \text{beam limited} \\ \dfrac{c\tau_p R_i \Phi_H}{2\sqrt{2}\ \sin\theta} , & \text{pulse limited} . \end{cases} \tag{40}$$

Here, c is the speed of light, τ_p is pulse width, and Φ_V and Φ_H are one-way, half-power, full-width antenna beamwidths in the vertical and horizontal. Then, to first order in U_o/C, the geometric part of the MTF is defined by

$$- it + r = \frac{1}{P_o} \frac{\partial P}{\partial (U_o/C)} \tag{41}$$

where the subscript 'o' indicates the zeroth order value, and one may show that the tilt term for vertical polarization is given by

$$t = + \cos\phi_o \tanh Kd \left[\frac{4\sin\theta_o \cos\theta_o}{1 + \sin^2\theta_o} + \frac{4\sin\theta_o}{\cos\theta_o + .111} - 4\tan\theta_o + \frac{1}{\psi_o} \left.\frac{\partial\psi}{\partial k}\right|_o (2k_o \cos\theta_o) + f(\theta_o) \right] . \tag{42}$$

The function $f(\theta_o)$ depends on illumination conditions and is given by

$$f(\theta_o) = \begin{cases} \tan\theta_o , & \text{beam limited} \\ -\cot\theta_o , & \text{pulse limited} . \end{cases} \tag{43}$$

Similarly, the tilt term for horizontal polarization is

$$t = + \cos\theta_o \tanh Kd \left[\frac{.444\sin\theta_o}{.111\cos\theta_o + 1} - 4\tan\theta_o + \right.$$

$$\left. + \frac{1}{\psi_o} \left.\frac{\partial\psi}{\partial k}\right|_o (2k_o\cos\theta_o) + f(\theta_o) \right] . \tag{44}$$

Finally, one finds that the range change term is given by

$$r = \begin{cases} \dfrac{2\tan Kd}{hK} , & \text{beam limited} \\ \dfrac{3\tan Kd}{hK} , & \text{pulse limited} . \end{cases} \tag{45}$$

Obviously, this term becomes less important as the altitude of the antenna increases. Figure 1 shows t as a function of incidence angle for both polarizations and for beam and pulse limited cases. Here we assume that $\phi_o = 0$, that we operate in deep water, and that $\psi(k) \propto k^{-4}$. Note that, in order to properly include second order effects, t and r given here should be multiplied by $P_o/\overline{P}$.

When removing these geometric effects from a measured MTF, one must be careful to account for the complex nature of m. In this respect, the phase of the MTF must be properly defined. Above, we indicated that when making MTF measurements at one point in space and computing cross-spectra in the frequency domain, the phase of the MTF is positive if it leads the wave crest and zero at the crest. In principle, though somewhat difficult in practice, another method of computing the MTF would be to measure received power and wave height at a single time as a function of space. Then the MTF would be computed using cross-spectra in the wavenumber domain. In many applications, it is this MTF as a function of long wave number which is required as we shall discuss below. If the standard convention of $\exp(-i2\pi ft)$ is used in defining temporal and spatial variation of surface waves, as was done for example in Eqs. 24 and 25, then a positive phase for a spatial MTF indicates that the received power lags the wave crest in space. Thus phases of the MTF have opposite signs depending on whether temporal or spatial variations of received power are being addressed.

5. APPLICATIONS OF THE MODULATION TRANSFER FUNCTION

The modulation transfer function enters into the theoretical interpretation of most active microwave radars which view the ocean surface. Here we shall review the manner in which the MTF impacts the operation of real and synthetic aperture imaging radars, wave spectrometers, scatterometers, and altimeters when they are used to sense the air/sea interface.

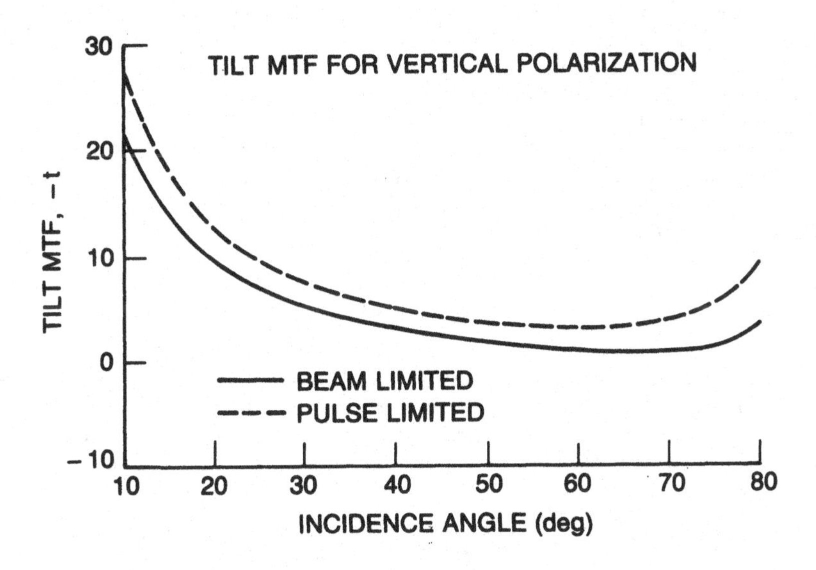

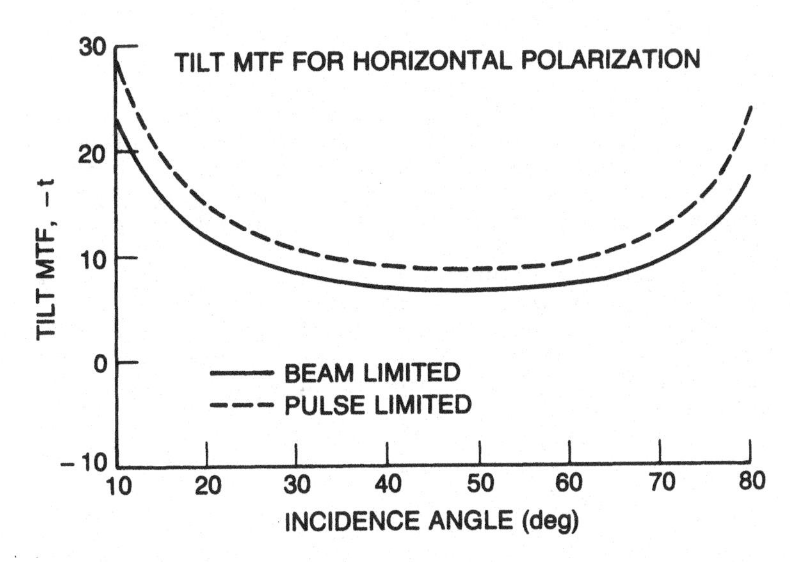

Figure 1. Tilt MTF for vertical and horizontal polarizations and for beam and pulse limited illuminated areas.

5.1 Real and Synthetic Aperture Imaging Radars

The image $I(x_o,y_o)$ produced by either a real or synthetic aperture radar system may be conveniently represented by

$$I(x_o,y_o) = \int \sigma_o(x,y)\rho(y-y_o)\Gamma(x-x_o)dxdy \tag{46}$$

where y is the range direction, x is the azimuth direction, $y_o = ct$ where t is time, and $x_o = Vt$ where V is aircraft velocity (Alpers et al., 1981). The function $\rho(y-y_o)$ represents the pulse shape of the transmitted signal; both imaging radars achieve their range resolution using narrow pulses.

For a real aperture radar (RAR), the function $\Gamma(x-x_o)$ is simply the antenna gain function; azimuthal resolution is achieved in real aperture imagery using a narrow antenna beamwidth. We may write Eq. (46) for this case as

$$I(x_o,y_o) = \bar{\sigma}_o(x_o,y_o)\ \Delta A(x_o,y_o) \tag{47}$$

where $\bar{\sigma}_o$ is the mean cross section over the resolution cell $\Delta A(x_o,y_o)$ which varies due to tilt and range change. Thus if ocean waves are viewed by a RAR, the image simply represents the spatial variation of received power and may be written, in a manner similar to Eqs. 20 - 23,

$$I(x_o,y_o) = \bar{P}\left[1 + \int m(K_x,K_y)\big(U(K_x,K_y)/C(K)\big)\ e^{i(K_x x_o + K_y y_o)}\ dK_x dK_y\right] \tag{48}$$

That is, intensity variations in real aperture imagery of ocean waves are due to the product of wave slope and the MTF as a function of long wave vector. Thus, if wave height information is to be extracted from RAR imagery, the MTF must be known for all long wavelengths and all angles between wave propagation direction and antenna pointing direction.

For a synthetic aperture radar (SAR), the situation is more complex (Plant and Keller, 1983; Hasselmann et al., 1985). The function $\rho(y-y_o)$ is still the pulse shape but $\Gamma(x-x_o)$ is now a normalized reflectivity variance spectrum, or normalized Doppler spectrum measured over a small area, with a Doppler frequency, f_d, related to azimuthal position by

$$f_d = \frac{k_o V(x-x_o)}{\pi R_o} - \frac{k_o v_r}{\pi} \tag{49}$$

where R_o is the range to the resolution cell at y_o and v_r is the radial velocity of the cell. Thus both the center position and the bandwidth of $\Gamma(x-x_o)$ may vary across the scene. If we assume that the bandwidth of this spectrum is very narrow and that it's center frequency depends on v_r, then we may write

$$\Gamma(x-x_o) = \Delta A\delta \left[\frac{k_o V(x-x_o)}{\pi R_o} - \frac{k_o v_r}{\pi}\right]$$
$$= \Delta A \left(\frac{\pi R_o}{k_o V}\right) \left(1 - \frac{R_o}{V}\frac{\partial v_r}{\partial x}\right) \delta(x-x_o) \tag{50}$$

where we have assumed that

$$\left| \frac{R_o}{V} \frac{\partial v_r}{\partial x} \right| << 1 \quad . \tag{51}$$

This formulation neglects the degradation of SAR azimuthal resolution caused by the finite bandwidth of Γ. Thus the image produced by a SAR may be represented by

$$I(x_o,y_o) = \bar{\sigma}_o(x_o,y_o)\Delta A(x_o,y_o) \left[\frac{\pi R_o}{k_o V}\left(1 - \frac{R_o}{V}\frac{\partial v_r}{\partial x}\right)\right] \quad . \tag{52}$$

This is just the RAR image multiplied by a factor which takes into account the effects of scatterer motion on the apparent azimuthal position of the scatterer in a SAR image. Once again, interpretation of a SAR image to obtain wave height information requires knowledge of the MTF as a function of long wavelength and direction. In cases where the bandwidth of Γ is not small or the approximation of Eq.(51) cannot be made, interpretation of a SAR image is even more complex and is not simply related to the MTF.

5.2. Wave spectrometers

The two most prominent types of non-imaging wave spectrometers are ROWS and TRIFAR. Both techniques operate using a conically-scanning, pencil-beam antenna. The azimuthal dimension of the footprint in both cases is large compared to the dominant surface wavelength so waves travelling at large angles to the range direction are not sensed by the system as discussed at the end of Section 3. Scanning the antenna azimuthally varies the range direction allowing one to determine the directionality of the wave pattern.

In the ROWS, or short pulse technique, range resolution is achieved through the use of the narrow pulse which sweeps across the footprint (Jackson, 1981). Thus fluctuations in received power are produced by range-travelling waves whose wavelengths are long compared to the pulse width. We may write the power received by a ROWS system in a manner similar to that of RAR and SAR imagery, Eq. 46. In this case, however, the function Γ is very broad so cross section variations in the azimuthal, or x direction are averaged out. Furthermore, since the antenna is scanned, the y direction is variable and should more properly be indicated by a range variable, say r'. Thus, the power received by ROWS is

$$P_r(r',\Phi_o) = \bar{P} \left[1 + \int m(K,\Phi_o)\ (U(K,\Phi_o)/C(K))\ e^{iKr'} dK\right] . \quad (53)$$

where Φ_o is the (variable) azimuthal pointing direction. Fourier transforming P_r and taking the modulous squared yields a measure of the wave spectrum in the direction Φ_o:

$$\left| \tilde{P}_r \right|^2 \propto \bar{P}^2 \left[\delta(K) + |m(K,\Phi_o)|^2 \left| \frac{U(K,\Phi_o)}{C(K)} \right|^2\right] . \quad (54)$$

Thus the MTF, as a function of wavenumber, need only be known in this case for waves travelling along the horizontal antenna look direction. This is a simpler situation than the case of RAR or SAR imagery where the MTF was required for all possible angles between the wave and antenna directions.

TRIFAR may be thought of as the Fourier transform of ROWS (Plant and Reeves, 1984). Instead of achieving high range resolution through a short pulse, TRIFAR transmits three frequencies spaced Δf apart. The beat wavelength $\Delta\lambda$ corresponding to Δf is adjusted to be resonant with a long surface wave. The mean received signal is caused only by this resonant surface wave. Scanning Δf and antenna azimuthal position then produces the complete wave spectrum. The signal produced by TRIFAR after processing may be written

$$S_{\Delta k} \propto \bar{P}^2 \ |m(K,\Phi_o)|^2 \left| \frac{U(K,\Phi_o)}{C(K)} \right|^2 . \quad (55)$$

Thus, like ROWS, only the MTF for waves propagating along the horizontal antenna look direction need be known in order to obtain wave spectra from TRIFAR.

5.3. Scatterometers

The MTF enters into scatterometry somewhat more subtly than is the case for RAR or SAR imagery or for wave spectrometry (Plant, 1986). As mentioned above, experiments have indicated that the mean received power from the sea surface depends on the mean squared dominant wave slope, a second order effect. In explaining this effect theoretically two factors must be considered. First the change in the mean spectral density of short, wind-generated waves caused by longer waves must be determined. Then the effects of surface tilt and its interaction with this hydrodynamic modulation must be determined to second order in long wave slope. Changes in the mean spectral density of the short waves due to their interaction with the long waves may be determined using the two-scale model and conservation of action. In this model, modulations in the short wave spectral density interact with the variable currents produced by the longer waves. In order to produce a change in the mean short wave spectral density, the hydrodynamic modulation must be 90° out of phase with the variable currents. This can be seen by considering an average of the action conservation equation:

$$\overline{\left[\frac{\partial}{\partial t} + (\vec{c}g + \vec{U}) \cdot \frac{\partial}{\partial \vec{x}} + \frac{\partial}{\partial \vec{x}} (\vec{k}\cdot\vec{U}) \cdot \frac{\partial}{\partial \vec{k}}\right] \cdot \left[\bar{N} \left(1 + \int m'(\vec{K}) \; (U(\vec{K})/C(K)) \; e^{i\vec{K}\cdot\vec{x}} d\vec{K}\right)\right]} = \bar{S}_o \tag{56}$$

where $\bar{N}$ is average short-wave action, $\vec{c}_g$ is group velocity, and S_o represents input and dissipation terms. In the steady state, $\partial N/\partial t$ = 0 and any remaining terms involving products of m′ and U contain spatial gradients which result in iK factors. Thus the average causes terms involving the real part of the hydrodynamic MTF, m′, to go to zero. Only terms involving the imaginary part of m′, ie, that part in phase with or 180° out of phase with long wave slope, play a role in changing the mean level of the received power.

In a similar manner, geometric effects of surface tilt enter into changing the mean received power to second order in long wave slope. In this case, the tilt modulation is always imaginary so it enters into the second order expressions either as a second order tilt term or as the product of the first order tilt term with the imaginary part of m′. The latter effect is the cause of the upwind/downwind asymmetry observed in scatterometer cross sections since when the direction of the antenna changes by 180° the tilt modulation also changes by 180° (See Eqs. 41 and 43) whereas m′ remains constant, that is, the short waves do not change their position on the long wave because the antenna changes direction.

Thus the MTF enters into scatterometry not as a single entity but as hydrodynamic and tilt modulations separately. Only the imaginary part of the hydrodynamic MTF needs to be known for the purpose of scatterometry but, once again, it must be known for all wave numbers and wave propagation directions in order to evaluate expressions for the mean cross section properly.

5.4 Altimeters

Jackson (1988) has recently examined the effect of short wave modulation on the sea state bias exhibited in altimeter measurements. This bias is the effect of sea state on the measurement of the distance from the altimeter to the mean sea surface. It is estimated that mean sea level as measured by an altimeter is lower than the true sea level by amounts approximately equal to 4 to 5% of the significant wave height. Jackson's expression for this bias, $\bar{\zeta}_{em}$, may be written in our notation as

$$\bar{\zeta}_{em} = - \frac{S_L^2}{S^2} \overline{\left(\frac{\partial \gamma}{\partial x}\right)^2 \gamma^*} - \frac{\overline{\nabla \gamma_s}^2}{S^2} \overline{(\nabla \gamma_s)^2 \gamma^*} \tag{57}$$

where the overbar indicates an average over the long waves, S_L^2 is mean squared long wave slope, S^2 is mean total squared wave slope, and

$$\nabla \gamma_s = \int k^2 \psi(\vec{k}) d\vec{k} \tag{58}$$

is the short wave mean squared slope. Since the short wave spectral density is modulated by the long waves, it is convenient to write

$$\psi(\vec{k}) = \overline{\psi(\vec{k})} \left[1 + \int m' \; (U/C) \; e^{i\vec{k}\cdot\vec{x}} \; d\vec{k}\right] \tag{59}$$

In taking the average in Eq. 57, only the part of Eq. 59 in phase with the long wave will survive. Thus we get for the case of a single long wave on the surface,

$$\bar{\zeta}_{em} = - \frac{\overline{\nabla \gamma_s}^4}{S^2} m'_r K a^2 \tag{60}$$

where m'_r is the real part of the hydrodynamic MTF. This shows that $\bar{\zeta}_{em}$ depends on both the real and imaginary parts of the hydrodynamic MTF since, as pointed out above, $\overline{\psi(\vec{k})}$ depends on the imaginary part. This latter dependence is relatively weak, however. Note that in this case, we must know m' not only as a function of long wave number and direction but for all short waves as well.

Table 1 shows the MTF applicable to each class of instrument along with the parameters of the long waves on which it depends. Note that the MTF measured by tower-mounted microwave systems can only be related to those required in instrument applications using a long wave dispersion relation.

6. LIMITATIONS OF THE MTF CONCEPT

The concept of a modulation transfer function relies explicitly on the assumption that perturbations in short wind wave equilibrium are weak enough that a first-order expansion in terms of the perturbing current is sufficient to account for changes in the short wave spectrum. Furthermore, surface tilts are assumed to be small enough that geometric effects on the microwave backscatter are linear in tilt angle. Experience indicates that these assumptions are fulfilled

TABLE 1

Modulation Transfer Functions Applicable to Different Instruments

Instrument	Applicable MTF*
Imaging Radars	$m(K_x,K_y)$
Wave Spectrometers	$m(K,0)$
Scatterometers	Imaginary part of $m'(K_x,K_y)$
Altimeters	$m'(K_x,K_y)$
Tower Radars	$m(f,\phi_o)$

* $m = m' - it + r$

reasonably well for many naturally occurring long surface waves. Mean levels of received power have been shown to depend on surface slope to second order, however, requiring some modification of the concept but still leaving it intact.

Of course, situations do occur on the ocean for which the concept of an MTF must break down. High sea states with significant whitecap coverage are an obvious example. Other examples might be ship wakes if the long waves generated by the ship are steep enough to break or large-scale current gradients where the perturbing influence exists for long enough time and space scales to change short wave spectral densities more strongly than a first order theory can explain. In such cases, one may require recourse to conservation of action equations to explain the full extent of the perturbing influence on short wave spectral densities.

In a wide variety of naturally-occurring oceanic conditions, however, the concept of the MTF is very useful in explaining features of signals received by active microwave systems viewing the surface. Thus efforts to understand the nature of the full MTF for all short wave numbers as a function of all long wave parameters and the causes of the observed modulation levels must proceed if we are to be able to properly interpret the output of these systems. Our present level of understanding is such that we cannot explain the large observed magnitudes of the MTF and do not know its behavior for all long wave parameters. The situation deserves attention.

7. REFERENCES

Alpers, W., D.B. Ross, and C.L. Rufenach, On the detectability of ocean surface waves by real and synthetic aperture radar, J. Geophys. Res., **86**, 6481-6498, 1981.

Feindt, F., J. Schroter, and W. Alpers, Measurement of the ocean wave-radar modulation transfer function at 35 GHz from a sea-based platform in the north sea, J. Geophys. Res., **91**, 9701-9708, 1986.

Hasselmann, K., R.K. Raney, W.J. Plant, W. Alpers, R.A. Shuchman, D.R. Lyzenga, C.L. Rufenach, and M.J. Tucker, Theory of synthetic aperture radar ocean imaging: A MARSEN view, J. Geophys. Res., **90**, 4659-4686, 1985.

Jackson, F.C., An analysis of short pulse and dual frequency radar techniques for measuring ocean wave spectra from satellites, Radio Sci., 16, 1385-1400, 1981.

Jackson, F.C., A two-scale model of the altimeter em bias, accepted by J. Geophys. Res., 1988.

Keller, W.C., and J.W. Wright, Microwave scattering and the straining of wind-generated waves, Radio Sci., **10**, 139-147, 1975.

Keller, W.C., W.J. Plant, and D.E. Weissman, The dependence of X band microwave sea return on atmospheric stability and sea state, J. Geophys. Res., **90**, 1019-1029, 1985.

Plant, W.J., W.C. Keller, and J.W. Wright, Modulation of coherent microwave backscatter by shoaling waves, J. Geophys. Res., **83**, 1347-1352, 1978.

Plant, W.J., and W.C. Keller, The two-scale radar wave probe and SAR imagery of the Ocean, J. Geophys. Res., **88**, 9776-9784, 1983.

Plant, W.J., W.C. Keller, and A. Cross, Parametric dependence of ocean wave-radar modulation transfer functions, J. Geophys. Res., **88**, 9747-9756, 1983.

Plant, W.J., and A.B. Reeves, The dual-frequency scatterometer re-examined, Proceedings of the URSI Commission F Symposium and Workshop, Shoresh, Israel, 191, May 14-23, 1984.

Plant, W.J., A two-scale model of short wind-generated waves and scatterometry, J. Geophys. Res., **91**, 10735-10749, 1986.

Schroter, J., F. Feindt, W. Alpers, and W.C. Keller, Measurements of the ocean wave-radar modulation transfer function at 4.3 GHz, J. Geophys. Res., **91**, 923-932, 1986.

Wright, J.W., W.J. Plant, W.C. Keller, and W.L. Jones, Ocean wave-radar modulation transfer functions from the west coast experiment, J. Geophys. Res., **85**, 4957-4966, 1980.

MEASUREMENT OF SHORT WAVE MODULATION BY LONG WAVES USING STEREOPHOTOGRAPHY AND A LASER-SLOPE SENSOR IN TOWARD

O.H. Shemdin

Ocean Research and Engineering
255 S. Marengo Avenue
Pasadena, California 91101
USA

ABSTRACT. The hydrodynamic modulation of short surface waves by swell is measured with stereophotography and a laser-slope sensor. The results are consistent with the relaxation model introduced by Keller and Wright (1975). A comparison with the modulation transfer function obtained with a tower-based L-Band radar indicates that hydrodynamic modulation is only a fraction of the total modulation. The conclusion arrived at in this paper is that the mechanisms involved in radar backscatter are understood at L-Band in low to moderate sea states. Work is in progress to achieve a similar understanding at X-Band.

1. INTRODUCTION

The modulation of short surface waves by long gravity waves is of interest in remote sensing of ocean surface waves with synthetic aperture radars (SAR). The early theoretical investigations of the hydrodynamic modulation of short waves by long waves were reported by Keller and Wright (1975) and Alpers and Hasselmann (1978). The theory, denoted as "Relaxation Theory", is discussed by Hwang and Shemdin (1988) in light of the recent TOWARD laser-slope measurements.

Direct measurement of the hydrodynamic modulation of short waves was attempted in the laboratory by Shemdin (1978) using a laser-slope sensor. The same measurement technique was used in the field during TOWARD. Hwang and Shemdin (1988) exercise care to account for the influence of orbital and ambient currents in converting the measurements from frequency to wave number domain. Satisfactory agreement is reported between the hydrodynamic modulations from the TOWARD measurements and

G. J. Komen and W. A. Oost (eds.), Radar Scattering from Modulated Wind Waves, 173–181.

theoretical predictions using the relaxation theory.

Direct measurement of short surface waves using stereophotography is demonstrated by Shemdin, Tran and Wu (1988) to yield direct measurement of the wave number spectrum. In TOWARD time-lapse stereophotography is used to determine the modulation in wave number spectra of short waves by the long waves. In this paper, the hydrodynamic modulation values obtained with stereophotography and the laser-optical sensor are compared with backscatter modulation obtained with the NRL L-Band radar to determine the relative importance of hydrodynamic modulation compared to the total radar backscatter modulation.

2. FIELD MEASUREMENTS

The TOWARD experiment was executed from the Naval Ocean Systems Center (NOSC) tower located offshore of Mission Beach, San Diego, during October 84-January 86. The geographic location of the tower is shown in Figure 1. An overview of the experiment is given by Shemdin (1988). Swells generated by storms in the North Pacific Ocean penetrate the Southern California Bight through the island window north of San Clemente Island and south of San Nicolas Island. The highly directional swell from 290 degrees, observed at the NOSC tower, allows delineation of the backscatter mechanisms from the sea surface. The measurement systems used to determine the modulation of short waves by the long waves are: stereophotography, laser-slope sensor and L-Band tower-based radar.

2.1 Stereophotography

A detailed description of this technique and its use for measuring short surface waves is given by Shemdin, Tran and Wu (1988). Briefly, the technique utilizes two cameras separated by a horizontal distance (base) deployed at an elevation (height) above the sea surface, as shown in Figure 2. This technique was used by Cote et al (1960) and Holthuijsen (1983) to obtain wave number spectra of long surface waves. The use of this technique to measure short surface waves has only recently been employed.

2.2. Laser - Slope Sensor

This instrument is described in detail by Shemdin (1980), Tang and Shemdin (1983) and by Shemdin and Hwang (1988). Briefly, it consists of a vertical laser beam deployed vertically below the water surface, with the refracted laser beam detected above the water surface by an optical receiver,as shown in Figure 3.

The modulation of short wave spectra by long waves is obtained by analyzing 1.0 sec time series. The location along the long wave profile is determined by conditional sampling. Modulated slope-frequency spectra are obtained by conditional sampling of 900-one second slope-frequency spectra.

2.3 Radar Backscatter Measurements

L-Band radar backscatter measurements were obtained during TOWARD by Plant and Keller (1988). Selected data sets were provided by the authors for further processing and comparison with the measurements obtained with stereophotography and the laser slope sensor.

3. ANALYSIS OF RESULTS

A comparison between the hydrodynamic modulations determined from the laser slope data and stereophotography, and the theoretically predicted modulations using the relaxation theory is shown in Figure 4. These results provide direct verification of the hydrodynamic modulation of short waves by swell under field conditions. The theoretical modulus of modulation, 4.0 based on k for the wave number spectrums, is within the data scatter.

Radar backscatter measurements, on the other hand, show the modulus of modulation to be as high as 15 with values reaching up to 30 (see Wright et al, 1980). Clearly, such magnitudes cannot be ascribed to hydrodynamic modulation alone. A critical review of radar backscatter from the sea surface is presented by Kasilingam and Shemdin (1988). Typical simulations for the moduli of modulation for hydrodynamic, tilt and range are shown in Figure 5. The hydrodynamic simulation is based on the relaxation theory. The tilt modulation is as described by Valenzuela (1978), and the range modulation is as described by Wright et al (1980). The simulation results shown in Figure 5 confirm the experimental observations that hydrodynamic modulation is only a fraction of the total radar backscatter modulation.

4. SUMMARY AND CONCLUSION

The results presented in this paper focus on modulation of 30 cm long waves, as only L-Band radar backscatter measurements are considered. The results clearly indicate that the relaxation theory does provide reasonable estimates of the hydrodynamic modulation of short waves by swell. The results presented have also shown that both hydrodynamic and tilt modulations are important contributors to the total radar backscatter modulation. Although these fingings are consistent with prevailing impressions, they provide the first direct verification of radar backscatter mechanisms at L-Band.

A careful validation of the mechanisms at X-Band and higher frequencies is planned for in the SAXON experiment, with field operations executed in September 1988.

ACKNOWLEDGEMENT

The TOWARD experiment was sponsored by the SAR Ocean Program of the Office of Naval Research with Commander Tom Nelson as Program Manager and Mr. Hans Dolezalek as Technical Moniter of the ocean SAR effort. Dr. William Plant provided the L-Band radar backscatter data.

REFERENCES

Alpers, W. and K. Hasselmann (1978) the two-frequency microwave technique for measuring ocean-wave spectra from an airplane or satellite, Bound. Layer Met., 13, 215-30.

Cote, L. F., J. O. Davis, W. Marks, R. F. McGough, E. Mehr, W. J. Pierson, Jr., J. F. Ropek, G. Stephenson and R. C. Vetter (1960) The Directional Spectrum of a Wind Generated Sea as Determined from Data Obtained by the Stereo Wave Observation Project, Meteorological Papers, Vol. 2, No. 6, New York University, 88 pgs.

Holthuijsen, L. H. (1983) Observations of the Directional Distribution of Ocean-Wave Energy in Fetch-Limited Conditions, J. Phys. Oceanogr., 13, 191-207.

Hwang, P. A. and O. H. Shemdin (1988) Modulation of Short Waves by Long Surface Waves, submitted to J. Fluid Mech.

Kasilingam, D. amd O. H. Shemdin (1988) Simulation of Tower-Based L-Band Radar Measurements in Toward, submitted to J. Geophys. Res.

Keller, W. C. and J. W. Wright (1975) Microwave scattering and the Straining of Wind-generated Waves, Radio Science., 10, 139-47.

Plant, W. J., W. E. Keller, and A. Cross (1983) Parametric Dependence of Ocean Wave-radar Modulation Transfer Function, J. Geophys. Res., 88, 9747-56.

Shemdin, O. H. (1978) Modulation of centimetric Waves by Long Gravity Waves: Progress Report on Field and Laboratory Results. In Turbulent Fluxed through the Sea Surface, Wave Dynamics, and Prediction, ed. A. Favre and K. Hasselmann, Plenum Publication, 235-55.

Shemdin, O. H. (1988) Tower Ocean Wave and Radar Dependence (TOWARD): An Overview, J. Geophys. Res., (in press).

Shemdin, O. H. and P. A. Hwang (1988) Comparison of Measured and Predicted Sea Surface Spectra of Short Waves, J. Geophys. Res., (in press).

Shemdin, O. H., S. Wu and M. Tran (1988) Directional Measurement of Short Waves with Stereophotography, J. Geophys. Res., (in press).

Tang, S. C. and O. H. Shemdin (1983) Measurement of High Frequency Waves Using a Wave Follower. J. Geophys. Res., 88, 9832-40.

Thompson, D. R. and R. F. Gasparovic (1986) Intensity Modulation in SAR Images of Internal Waves, Nature. 320, 345-58.

Valenzuela, G. R. (1978) Theories for the Interaction of Electromagnetic and Ocean Waves - A Review, Bound. Layer Meteo., 13, 61-85.

Wright, J. W., W. J. Plant, W. C. Keller and W. L. Jones (1980) Ocean Wave-radar Modulation Transfer Functions from the West Coast Experiment, J. Geophys. Res., 85, 4957-66.

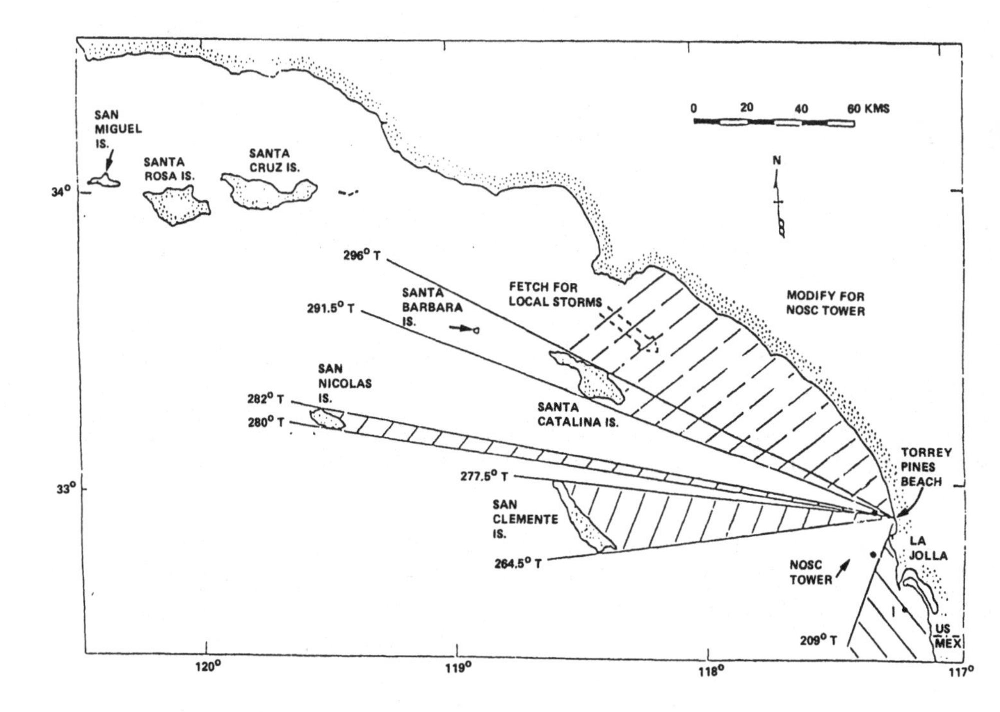

Figure 1. Southern California Bight, Showing Location of NOSC Tower and Windows of Wave Approach from the North Pacific.

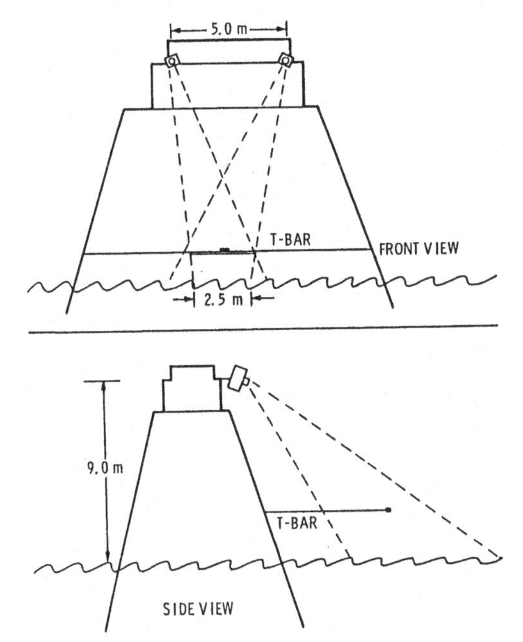

Figure 2. Schematic of Stereo-Geometry.

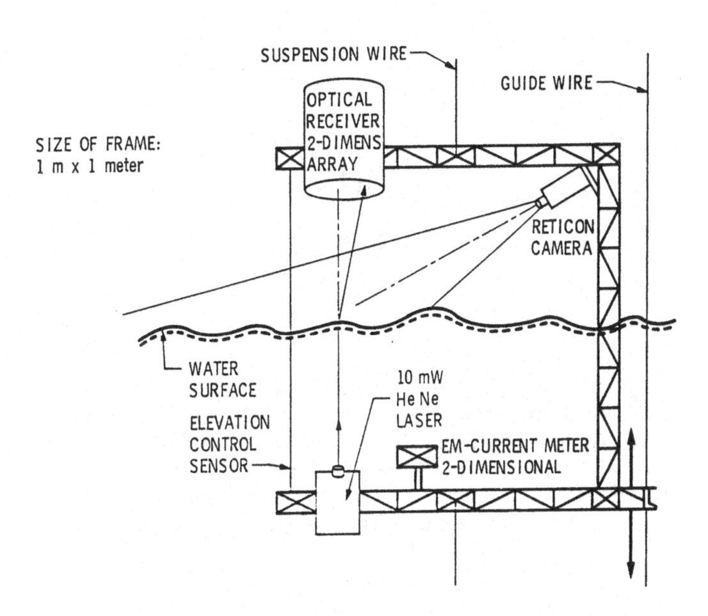

Figure 3. Wave Follower Instrument Frame.

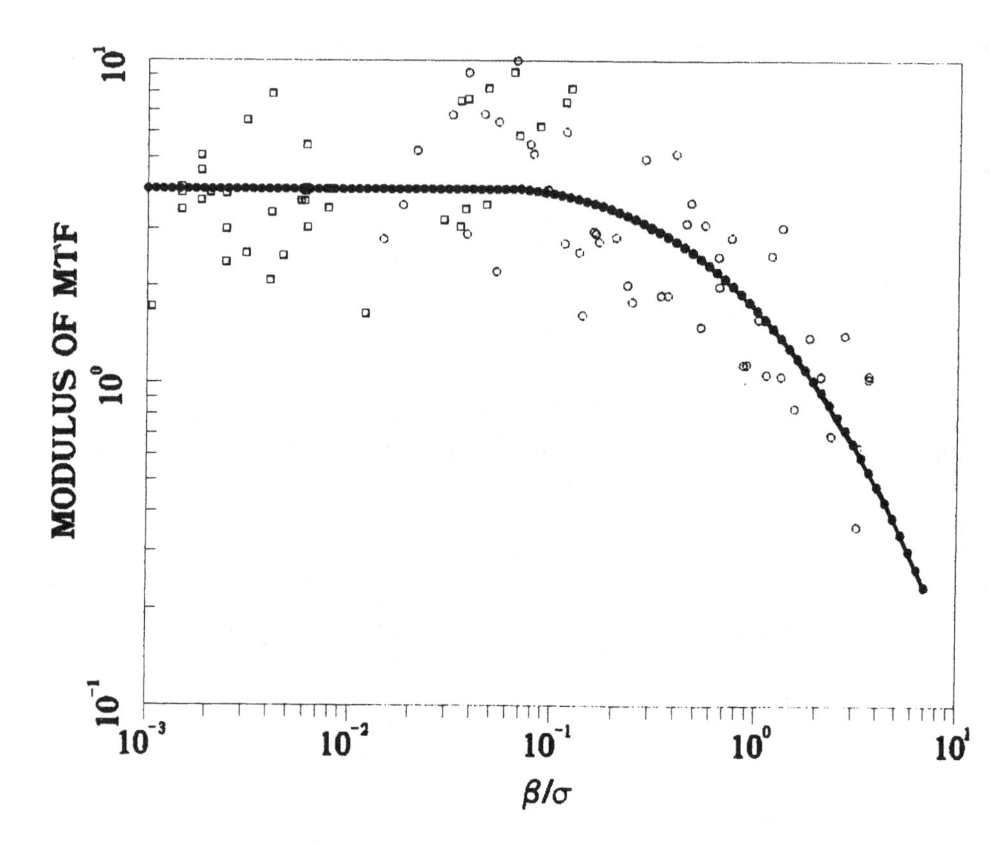

Figure 4. Comparison of Measured Hydrodynamic Modulus of Modulation with Relaxation Theory Predictions. β is Atmospheric Growth Rate and σ Frequency of Long Waves. —— Relaxation Model for β_r/β =2, where β_r is Relaxation Rate. □ -Stereo Data, ○ -Laser-Slope Data.

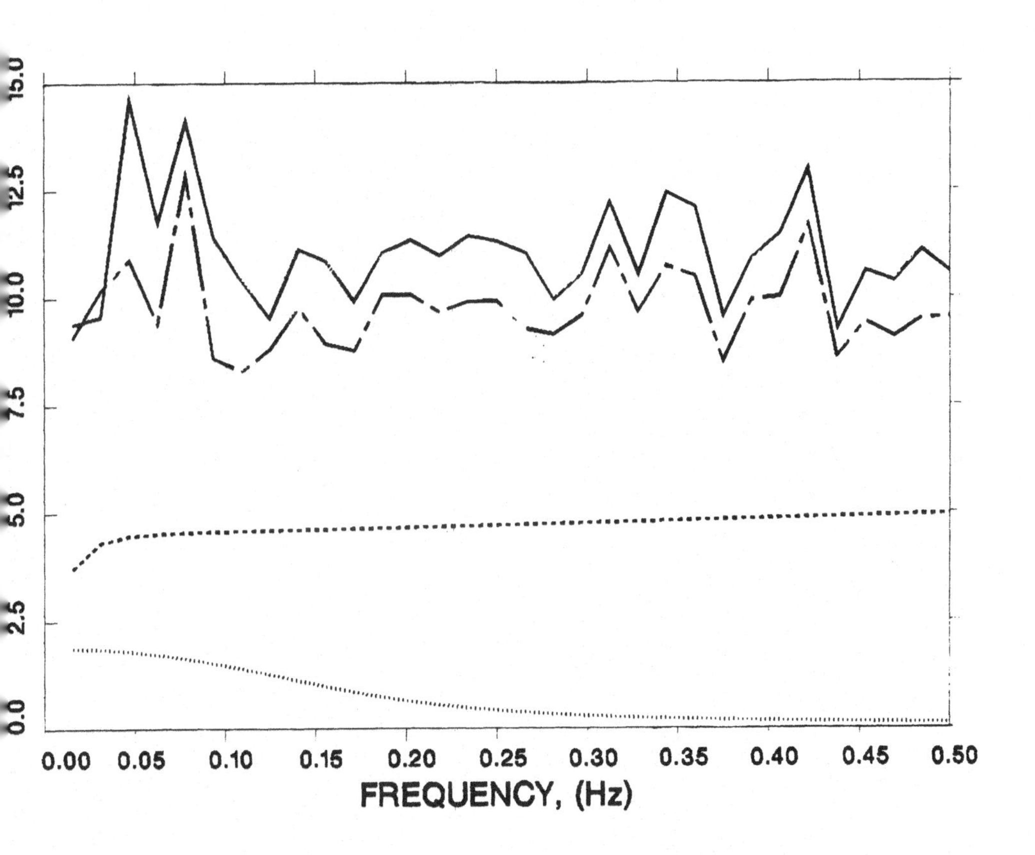

Figure 5. Simulated Hydrodynamic, Tilt and Range Components of Modulus of Modulation. ——Total MTF, —·—Tilt MTF, - - - Hydrodynamic MTF, ········ Range MTF.

THE SURFACE SIGNATURES OF INTERNAL WAVES IN THE OCEAN: INTRODUCTION AND SYSTEMATICS

J.O. Thomas and A.C. Edwards
Oxford Computer Services Ltd
52 Saint Giles'
Oxford, OX1 3LU,
U.K.

ABSTRACT. A brief description is given of work in progress on the modelling of the interaction between a surface current and the background surface wave field. The emphasis initially was on surface currents induced by internal waves but the techniques involved are sufficiently general for them to apply to any current which can be represented satisfactorily by piecewise linear segmentation over the straining region. Thus tidal flows over bathymetric features such as ridges and gulleys can be accommodated. The modelling is currently being extended to apply to shearing currents for comparison with observations of fronts and eddies.

INTRODUCTION

Space oceanography increasingly occupies a major place in the programmes of numerous research and other agencies throughout the world concerned with gathering new data on the global seas and with ocean monitoring and surveillance. The ERS-1 imaging radar which will orbit the earth as part of the European Space Agency's programme will add significantly to the data already available from the Seasat, Space Shuttle and other missions. The participation of the Ocean Sciences Division, ARE, Portland in collaboration with the Royal Aircraft Establishment (RAE), Farnborough, in the SIR-B and SIR-C and related projects provides UK Agencies and scientists with a prime opportunity in relatively new fields of ocean observational and measurement techniques. The remote sensing of ocean colour, sunglint monitoring from space, sea surface temperature measurements, wind and wave scatterometry and radar altimetry will provide new opportunities to study the world ocean and its dynamics with important consequences for understanding ocean/atmosphere interactions and for improving ocean and weather forecasting.

Radar observations of the surface signatures of internal waves may provide a method for deriving information about the deep interior of the ocean from satellites and are therefore of considerable importance to oceanographers and others concerned with ocean surveillance and monitoring. The presence of internal waves disturbs the position of the thermocline and therefore affects acoustic sound channels in

G. J. Komen and W. A. Oost (eds.), Radar Scattering from Modulated Wind Waves, 183–193.

the sea, with important consequences for sonic underwater communications and ocean surveillance. Considerable potential exists for the use of remote sensing techniques in the measurement of ambient acoustic background noise at the sea surface due to wind, rain, breaking waves etc. and as a contributor of extensive data to ocean modelling and forecasting programmes.

Oxford Computer Services Ltd (OCS) has recently completed a number of projects in the field of remote sensing of the ocean: these include, work on microwave radar imaging and on the detection and analysis of internal wave surface signatures and of surface gravity waves. The projects have been described in a series of OCS Reports, see also Bagg and Thomas 1984 and Bagg et al 1986.

The impact of the experience gained through the Seasat and the SIR-A and B experiments etc. should have a major influence on the UK contribution to the ESA programme involving the first European remote sensing satellite, ERS-1. Future space oceanography satellite experiments in the Columbus Space Station and Polar Platform programmes currently under discussion will undoubtedly also benefit from such work.

AIRCRAFT AND SATELLITE-BORNE OCEAN REMOTE SENSING

In our research in this area the main emphasis hitherto has been on understanding the radar signatures of internal waves as observed by synthetic aperture radar (SAR). Other oceanic surface roughness signatures of interest such as those of fronts, currents, bottom topography variations, eddies, oil slicks and ship wakes are generically related and are readily incorporated into the programme of data analysis. It is becoming clear that single instrument-single pass data have somewhat limited application in many areas of interest to our work. Co-located, simultaneous wind data, for example, are required for internal wave signature analysis. Composite SAR images are also desirable if not essential for such work.

In global terms it appears that the advances in theoretical signature modelling have not been matched by substantial programmes of image data analysis and by in-situ experiments. A major feature of the planning behind our present approach is that it envisages more emphasis <u>on applying</u> the theoretical work already developed to a variety of internal wave and other relevant data which have been accumulating from various sources (INTRADAN airborne radar videos, shuttle astronaut photographs, Nimbus 7 (sunglint) photographs etc.) while at the same time allowing continuing development of new ideas on the theory side. In particular the OCS surface wave - current interactions kinematics program hopefully will be applied to actual data. A necessary concomittant of this idea is the proposal to continue to evaluate cost-effective basic peripheral equipment requirements for image analysis. All this is required if ambitions to derive density stratification parameters on a routine basis from satellite imagery are to be realized.

RADAR SIGNATURES OF INTERNAL WAVES : FORWARD PROBLEM

Two problems are extremely important here. The first is to allow for the effect of the partial re-distribution of the wave energy of breaking long wavelength waves (i.e. long with respect to the Bragg wavelength) which can act as an isotropic source of wave energy for shorter wavelength waves - (this occurs in addition to the well known wave-wave interaction mechanisms (van Gastel 1987)). The inclusion of such a term would convert the Phillips (1984) wave action evolution equation into an integro-differential equation with potentially interesting solutions. Such an approach, while maintaining the fundamental importance of the Bragg mechanism, would begin an attack which would bring in effects not only at the Bragg wave-number k_B, but at multiples of k_B, at a band around k_B (cf Holliday 1987), and over a wide band with $k<k_B$. Such an approach may well account for some of the discrepancies from the Bragg mechanism increasingly voiced in the literature - in particular that the simple Bragg mechanism cannot explain the fact that SAR signatures of internal waves are clearly obtained even when the internal waves are viewed parallel to their 'crests'.

Secondly, present theories often involve unwarranted and unnecessary simplifications. It is suggested that precise 'hands-off' modelling is possible via the Phillips theory and that such an approach is a natural and highly desirable extension of current research. Thus an ultimate goal of our research is to open up the possibilities for thermocline/water density stratification studies - the inverse problem. The implication here is that if the surface current can be deduced from the radar signature, the inverse problem of obtaining the density profile with depth may be amenable to solution possibly without having to make gross assumptions about the nature of the solitons or the layered nature of the ocean.

Hand-in-hand with such theoretical achievements would go the development of algorithms for operational applications specifically aimed at the exploitation of ERS-1 and other data for water density parameterization (Thomas, Edwards and Perry 1986). Similarly there would be an immediate and continuous activity to accommodate the increasing availability of relatively low cost equipment and software for image analysis on the macro- and micro- levels.

In terms of the fundamental theory of electromagnetic wave scattering by rough surfaces illuminated by radar many problems such as multiple scatter and de-polarization require further investigation. A basic need is an evaluation of theory and experiment in relation to the wind dependence of the radar backscatter coefficient σ_o and the determination of the dependences of σ_o on radar frequency (particularly at C, L and X bands), and of the effects of depolarization etc. Recent new work carried out in OCS by Jeynes in this field is relevant and is described briefly in an accompanying paper in this volume. Such studies would have a particular importance for surface film (oil slick etc.) studies.

THE OCS HYDRODYNAMIC INTERACTION MODEL 'OCSHIM'

The theoretical basis for the modelling of radar surface roughness signatures associated with currents such as those induced by internal waves, tidal flow over bathymetry etc. is discussed in detail elsewhere and provides the theoretical framework for the sets of programs that constitute the Oxford Computer Services Hydrodynamic Interaction Model (OCSHIM) (Edwards 1988). These have been constructed over the past three years for the purpose of calculating the modulation and signature of short gravity wind waves on an ocean surface that is being affected by the action of a straining surface current - due, for example, to an internal gravity wave. The models are applicable to line-like surface features observed by, for example, imaging radars.

The specification of a steady state, one-dimensional surface current distribution enables the first half of OCSHIM (known as MAKE_DF) to compute and store all the refracting wave group paths that, when observed, have a specified wavevector - the principal, or Bragg, wavevector. The specification of environmental parameters then enables the second half of OCSHIM (known as USE_DF) to compute the expected radar backscatter signature at the principal wavevector by using the stored wave path data and by assuming that the Bragg mechanism is operative. (More complex backscatter mechanisms may be investigated by multiple runs of OCSHIM.)

1. Input parameters for the OCSHIM wave path generator MAKE_DF

1.1 The principal (or Bragg) wavevector

This is determined by the user inputting:-

a) The radar operating wavelength
b) The angle of incidence of the radar beam to the sea surface
c) The angle between the radar look direction and the normal to the observed imaged feature (assumed linear). The feature referred to corresponds to the image grey level changes resulting from the changes in σ_o arising from the changes induced in the sea surface background wave spectrum by the interaction with a surface current such as that induced by an internal wave.

1.2 The steady state, 1-D surface current distribution:

Several choices are available:-

a) An arbitrary user-defined surface strain rate distribution
b) An arbitrary user-defined sea depth distribution and a value for a constant tidal or current volume flux
c) Benjamin-Ono solitary wave(s)
d) Korteweg-de Vries solitary wave(s)
e) Joseph solitary wave(s)
f) Swell
g) Linear (sinusoidal) internal gravity waves

1.3 The linear feature dimensions and zoning requirements:

These are fixed by the user inputting:-

a) One-dimensional coordinate values that specify the boundaries in the direction normal to the linear feature
b) The number of zones that are required and into which the linear feature will be uniformly divided between the boundaries.

2. Input parameters for the OCSHIM signature generator USE_DF

2.1 Wind wave action input model:

Plant's (1982) formula and its generalization is used for the wind wave input model. Three parameters are required:-

a) The constant in Plant's formula (angular factors are ignored)
b) The value of the wind friction speed
c) The frequency dependence of the (generalised) Plant formula

2.2 Model of dissipation by wave breaking:

Phillips (1984) formulation is used. The one parameter required is:-

a) The Phillips wave breaking power exponent

2.3 Model of surface wave energy spectrum:

The simple Phillips power law spectrum is used. The one parameter required is:-

a) The wave number power exponent, p, of the surface wave <u>energy</u> spectral density written as $|k|^{-p}$ (where k is the wavenumber).

3. Description of MAKE_DF

The wave path generator, MAKE_DF, accepts as data input a principal wavevector, a steady state 1-D surface current distribution and the dimensions and zoning requirements of the linear feature. The program first sorts out the zone positions and calculates the current at each zone boundary. At each assumed point of observation (the zone boundaries) it is supposed that the 'visible' wave groups are those with the principal (Bragg and anti-Bragg) wavevector(s). These (two) have their apparent wave frequency calculated (on the assumption that the deep water short wavelength linear gravity wave dispersion relation holds everywhere) and the wave group wave paths are then traced back through the surface strain field of the linear feature to the point where they enter the straining region. At each point (zone boundary) along the wave path the intrinsic wave frequency is recorded. The program then sorts through the wave path frequencies so as to determine the wave path and stores the necessary data on a disk file. The program then moves on to the next observation point (zone boundary) and repeats the process until all the wave paths of the relevant wavegroups (with principal wavevector at the observation points equal to the Bragg and anti-Bragg wavevectors) are stored on disk.

4. Description of USE_DF

The signature generator, USE_DF, accepts as data input the stored wave-path data, parameters describing the wind wave input, a wave breaking dissipation parameter and a surface wave energy spectrum parameter. The signature (the ratio of the wave action spectral density to its equilibrium value) at each observation point is then constructed by means of piecewise analytic solutions to the wave action evolution equation. After the signature is formed at every observation point for the principal wavevector (which is the same as supposing that the Bragg mechanism is responsible for signature formation) then the program outputs the whole signature in graphical form and returns for the re-input of new environmental parameters.

THE WAVE ACTION EVOLUTION EQUATION

The wave action evolution equation takes the form

$$\dot{N} = \dot{Q}$$

where the left-hand-side denotes the rate of change with time of the wave action spectral density N. The right-hand-side is not to be thought of as the specific rate of change of a single entity Q : rather $\dot{Q}$ embodies the changes associated with processes which input and dissipate wave action. These include the input of N from wind-induced waves and dissipation by wave breaking. Other processes which contribute to $\dot{Q}$ include wave-wave interactions and the re-distribution of wave energy into parasitic capillaries etc. In our approach we have assumed that these latter two effects can be ommitted, at any rate, for L-band SAR imagery. Recently van Gastel (1987) has shown that it is important to include the effects of wave wave interactions at sub-centimetric wavelengths.

The solution for $\dot{Q}$ zero (sometimes referred to as the adiabatic case) is readily obtained. More interesting are the problems associated with getting solutions for the non-adiabatic case i.e. for $\dot{Q} \neq 0$. The form taken by the left-hand side of the equation is broadly understood and will not be discussed further, though, in passing, it is interesting to ask whether wave-wave interactions are best accommodated on the left- rather than the right-hand side of the wave action evolution equation.

Our approach to modelling the hydrodynamic interaction between surface current and surface waves is based on the Phillips formulation of the source-sink terms on the right-hand side. Though we assume the Bragg scattering process, the method is also amenable to the adoption of other mechanisms. Though we employ the wave action spectral density $N(\underline{k},\underline{x},t)$ where $\underline{x}$, $\underline{k}$ denote the spacial position, the wave vector of the short gravity wave under consideration, and t denotes time, it is the wave-height spectral density, ψ, that is relevant in the SAR detection of surface roughness changes and these are related through

$$N(\underline{k},\underline{x},t) = c(\underline{k})\,\psi(\underline{k},\underline{x},t)$$

where c is the phase speed of the short gravity wave involved. Though we may interchange N and ψ for the short gravity waves of interest herein, it is strictly not permissible to do this when the effect of surface tension is included in the dispersion relation, as is necessary when considering shorter (sub-centimetre wavelength) waves. Also excluded are non-linear effects due to large wave amplitudes.

A study of relevant wave kinematics over the straining region is central to our method which involves following a wave group. The apparent frequency Ω and the y component of $\underline{k}$ are constants of the motion where w is the intrinsic frequency and

$$\Omega = w + \underline{k}.\underline{U} \quad \underline{k}^2 = k_x^2 + k_y^2 \quad w^2 = g|k|$$

Here the current $\underline{U} = \underline{i}\, U_x(x)$ where the 0x axis is taken as the normal to the linear feature. The 0x axis is itself inclined at an angle γ to the radar look direction. The quantity w is the intrinsic wave frequency and g the acceleration of gravity. For short gravity waves on deep water the dispersion relation is as shown. Considerable progress can be made (without recourse to complicated ray-tracing which requires large scale computing facilities) by considering a systematic evaluation of the kinematics via representations in $\Omega_v w$ space and in $\Omega_v x$ space.

In general, to solve the $\dot{N}$ equation two routes have been followed:

1. The wave group method (e.g. Hughes 1978, Edwards 1985, Thomson and Gasparovic 1986)

2. The spectrum development method (e.g. Phillips 1984, Perry 1985).

In the former a particular wave group is followed through the straining region (the S-region where S is the straining rate $\partial_x U$) and $\underline{k}$ and N are both recorded as functions of x

$$N \equiv N[\underline{k}(\underline{x}),\ \underline{x}]$$

In the second route N is considered regardless of any particular group so that

$$N \equiv N(\underline{k},\underline{x})$$

where $\underline{k}$ and $\underline{x}$ are independent and comprise a 4-D space.
In the wave group method

$$d/dt = \partial/\partial t + \dot{\underline{x}}.\nabla$$

where $\dot{\underline{x}}$ is the space velocity of the wave group.
This method has certain analytical advantages, particularly through the use of the notion of modified wave action as suggested by Edwards 1985.

The spectrum development method has

$$d/dt = \partial/\partial t + \dot{\underline{x}} . \partial_{\underline{x}} + \dot{\underline{k}} . \partial_{\underline{k}}$$

$\dot{\underline{k}}$ corresponds to the "velocity" of a wave group in the fixed $\underline{k}$ space given by

$$\dot{\underline{k}} = - (\partial_{\underline{x}} U) . \underline{k}$$

This gives solutions at a fixed $\underline{k}$, for example $\underline{k} = k_B$ where k_B denotes the Bragg wave number. The quantities $\partial_{\underline{x}}$ and $\partial_{\underline{k}}$ denote the "space" derivatives of the 4-D space operating only on their respective coordinates.

In both approaches

$$\partial/\partial t = 0$$

The wave kinematics following wave groups across the S-region leads to

$$\Omega = w + G(w^4 - w_y^4)^{½} \quad \text{where } G = \sum |U(x)|/g$$

and $\sum = \pm$ according as k_{xB} (the x-component of the Bragg wave) is parallel or anti-parallel to U_x. This equation is referred to as the kinematic quartic (KQ).

An expression for the x-component of the space velocity, C_x+U, of the wave group is readily derived in terms of Ω, w, w_y for both the possible signs of k_x with respect to the x-axis. The quantity C_x is the x-component of the wave group velocity $\underline{C}$. This in turn leads to an evaluation of points on which the group space velocity $\dot{\underline{x}}$ is zero: the turning point quartic (TPQ). Note that for $\gamma=0°$, $w_y=0$ and the KQ becomes a quadratic.

The KQ enables us to compute the path history of a wave group (i.e. $k_x(x)$ and w(x)). Reflection points are given by the TPQ. The kinds of wave group paths and motions that can occur in practice in the S-region are conveniently classified in terms of "sectors" within each of which the behaviour of the wave group is analysed in terms of $\Omega_v x$ space or $k_v x$ space.

In considering $\dot{Q}$, an extremely important factor is the $\underline{k}$ dependent wind-wave relaxation rate, and its associated relaxation time $1/\mu$ (Alpers 1983, Alpers and Hennings 1984). We write

$$\dot{Q} = [\mu/(n-1)]N(1-(N/N_o)^{n-1})$$

where $N \equiv N(\underline{k})$ and $N_o = N_o(\underline{k})$. The number n (Phillips 1984) lies between 3 and 5 (not necessarily integer). In terms of the systematics of the problem note that n=2 corresponds to the form used by Hughes (1978)*

---

* Hughes's expression for μ [$\equiv$ his β_*] is widely used in recent American work (e.g. Thomson and Gasparovic 1986, Holliday et al 1986) despite Hughes's comment that its "specification is still a matter of some guesswork".

and by Shuchman et al (1985). Note also that taking a small perturbation $\delta N = N - N_o$ and retaining only first order terms in N leads to the results obtained by Alpers and Hennings (1984).

The relative degree of saturation, b, introduced by Phillips has provided a major step in pointing the way to solutions of the wave action evolution equation. Applied by him to the case of a current of the form $U_o f(\underline{x}/L)$ where L is a characteristic length, it provided the inspiration for more refined applications, for example to internal waves and flow over bathymetry. At this juncture our work departs from that of Phillips through allowance for the motion of the current i.e. the introduction of a current of the form $U_o(f[(x-vt)/L])$. This led to the important delineation of the circumstances in which it is permissible to ignore terms in $\partial b/\partial k$ and the realization of its general importance in the study of internal waves.

SOLUTION VIA THE CONCEPT OF PASSAGE

The evolution equation may be written

$$(C_x + U)\partial_x N = \left(\frac{\mu}{n-1}\right) N\left(1-\left(\frac{N}{N_o}\right)^{n-1}\right)$$

It is convenient to define a new independent variable, τ, particular to each wave group through

$$d\tau = dx \cdot \frac{\mu}{C_x+U} = \left(\frac{\mu}{-S}\right)\cdot \frac{dk_x}{k_x}$$

The quantity τ has been called the 'passage' (Edwards 1985). It leads under certain circumstances to analytical solutions in situations where S is a constant. In practice the flow in the straining region may be divided into a series of strips or zones in each of which S is constant. This piecewise approach can be made as accurate as we like by making the zones narrow enough. The formal solution may be written in the simple form

$$\left[Me^{\tau}\right]_{\tau_2}^{\tau_1} = \int_{\tau_2}^{\tau_1} M_o e^{\tau}\, d\tau$$

where τ_1 and τ_2 are arbitrary upper and lower limits of the passage along the wave group of interest. The modified wave action M is defined as

$$M = N^{-(n-1)} \qquad M_o = N_o^{-(n-1)}$$

M_o is the modified wave action outside the straining region and is calculated on the basis of an equilibrium (Phillips) spectrum.

It is interesting to note that Hughes (1978) defined a quantity called the "quiescence". This is his equivalent to our M and equals 1/N for n=2.

The concept of passage provides a powerful tool not only in the physical understanding of what is taking place over the S-region but because it lends itself to numerical and very often, in suitable circumstances, analytical solutions.

The OCSHIM codes have been used to produce typical signatures for the $\gamma=0°$ case as well as for situations when $\gamma\neq 0°$ for representative ocean environment parameters and internal wave currents. The approach to the construction of the codes has been modular. Work is in progress to develop various options concerning the scatter process, the form of the current and of the detailed nature of Q. In particular the method is being extended to include shearing currents. Perhaps, however, the most important feature of the work completed to date arises from the opportunity which our approach has presented for assessing the way in which various approximations to the solution which have been published relate to the general solution and for discussing the circumstances in which the approximate solutions are valid. One particular consequence has been the identification of the need to properly allow for the effects of wave refraction when considering the signatures that arise in low wind and/or high straining rate conditions. This is discussed in a companion paper which follows.

REFERENCES

Alpers,W.R.,1983. In *'Satellite microwave remote sensing'*(ed.T.D.Allen) Ellis Horwood, Chichester.

Alpers,W.R. & Hennings,I., 1984. J.Geophys.Res.,89,10529.

Bagg,M.T. & Thomas,J.O., 1984. Int.J.Rem.Sens., 5, 969.

Bagg,M.T., Edwards,A.C., Perry,J.R., Scott,J.C., Stacey,J.A.& Thomas,J.O. 1986. *'The SIR-B Mission: towards an understanding of internal waves in the ocean'* ARE Tech.Report No.86122.

Edwards,A.C., 1985. *'Modelling of selected SAR features induced by surface current variations'* OCS Report.

Edwards,A.C., 1988. *'Radar signature formation and the OCS Hydrodynamic Interaction Model OCSHIM'* OCS Report.

van Gastel, K., 1987. J. Geophys.Res.,92,11857-11865.

Holliday,D., St-Cyr,G., & Woods,N.E., 1986. Int.J.Rem.Sens.,7,1809-1834.

Hughes,B.A., 1978. J.Geophys.Res.,83, 455.

Perry,J.R., 1985. *'The imaging by SAR of internal waves and related ocean dynamics.'* OCS Report.

Phillips,O.M., 1984, J.Phys.Oceanog. 14, 1425-1433.

Plant,W.J., 1982. J.Geophys.Res., 87, 1961-1967.

Shuchman,R.A., Lyzenga,D.R. & Meadows,G.A., 1985. Int.J.Rem.Sens., 6, 1179.

Thomas,J.O., Edwards,A.C. & Perry,J.R., 1986. *'Evaluation of the operational implementation of internal wave analysis procedures with reference to ERS-1,'* OCS Report.

Thompson,D.R. & Gasparovic,R.F., 1986. Nature, 320, 345.

Acknowledgement: This work was funded by the Procurement Executive, MOD. We thank Dr J.C. Scott, ARE Portland, for his support.

THE SURFACE SIGNATURES OF INTERNAL WAVES IN THE OCEAN: SOME PARTICULAR CASES

A.C. Edwards and J.O. Thomas
Oxford Computer Services Ltd
52 Saint Giles'
Oxford, OX1 3LU,
U.K.

ABSTRACT. We present briefly an overview of work concerned with the inclusion of wave refraction effects in computing the radar surface roughness signatures of internal waves. This requires a discussion of wave kinematics in the straining region. It is suggested that important progress can be made in understanding the physics of the straining process by employing a system based on the use of 'sectors' and of representations in the $\Omega_v w$ and $k_y x$ planes. It is shown that in the special case of low winds so that the associated time-scales are relatively long, the refraction effects are especially important and can lead to the presence of fine structure in the theoretically modelled signatures. Some consequences such as the occurrence of wave trapping are described and the implications for experiments are outlined.

INTRODUCTION

The kinematic quartic equation may be written in the form

$$(a - z)^2 = z^4 - b^4$$

where

$$z = Gw \qquad a = G\Omega \qquad b = Gw_y \quad \text{and } G = \Sigma |U|/g.$$

The symbol Σ is ±1 according as the wave vector of interest (e.g. the Bragg vector k_{xS}) is parallel or anti-parallel to $\underline{U} = \underline{i}U_x$ where $\underline{i}$ is the unit vector along the 0x axis which is inclined at an angle $\gamma°$ to the radar range direction and $\underline{U}$ is the current. Ω and w are the apparent and intrinsic wave frequencies and g is the acceleration of gravity. The physical roots of the kinematic quartic obey $|z| \geq |b|$ and $a > z$.

For the wave group motion it is easy to show that the locus of points on which the space velocity $C_x + U_x$ is zero ($\underline{C}$ is the wave group velocity) is defined by

G. J. Komen and W. A. Oost (eds.), Radar Scattering from Modulated Wind Waves, 195–199.

$$z^4 - b^4 + 2z^3(a - z) = 0$$

This equation gives the turning points of the wave group and is referred to as the turning point quartic. The kinematic quartic may be used to trace the wave path and wave properties of the wave vector of interest. A discussion of the roots of the equation is beyond the scope of this brief overview but the Table reproduced below illustrates a convenient way, - in terms of "sectors" - which enables a systematic organization of the many variables involved. Wave groups within a particular sector behave in a way characteristic of that particular sector or domain. Five sectors are defined in all. In sector A the vectors k_{xs} and U_x are parallel. In sectors B and C, they are anti-parallel. In sector B the x-component of the wave vector is anti-parallel to both that of the current and the space velocity. It is necessary to divide B into B1, B2 and B3 as shown in the table. In B1,Ω is negative whereas in B2 and B3 it is positive. When two turning points exist, there are situations when three real roots are possible for the kinematic quartic. Two correspond to sectors C and B2 but the third root corresponds to sector B3.

Sector	Ω	Σ	σ	ntps
A	+	+1	+1	0
B1	-	-1	-1	0
B2	+	-1	-1	1 or 2
B3	+	-1	-1	0 or 2
C	+	-1	+1	1 or 2

Table showing sector properties. The last column gives the number of turning points, ntps, along the wave group path being followed through the straining region. The quantity σ is ± 1 according as the x-component of the group space velocity is parallel or anti-parallel to k_{xS}.

In the rest frame of a soliton travelling in the x direction with phase speed, V, and current amplitude U_a the effective current in the straining region is

$$U(x) = V - U_a.f(x)$$

and is +ve in the far field away from the straining region. f(x) corresponds to a sech^2 function for a Korteweg de Vries soliton. For illustrative purposes we choose f(x) to be triangular in shape over the S- region so that the strain rate is $\pm S$ where S is a positive constant. In $k_v x$ space in which wave refraction occurs it is possible to define a critical wavenumber k_{BC} for the B and C sector wave groups which are refracted to higher and lower values of w respectively. The critical wavenumber is found by equating the far field current V with a critical wave group speed C_{BC} giving a critical intrinsic wave frequency, w_{BC}, and a critical apparent wave frequency Ω_{BC} given, for the $\gamma=0°$ case and for V the same on each side of the S-region, by

$$w_{BC} = g/2V \qquad \Omega_{BC} = g/4V$$

For $0<w<g/2V$ we have C sector groups which enter the S-region from above and for $g/2V<w<g/V$ we have B2 sector wave groups entering the S-region from below. The wave path curvatures in the S-region in $w_v x$ space are opposite for B2 and C groups as may be seen from examining the equation

$$\frac{d\ln|k_x|}{dt} = -S$$

As the frequency w_{BC} is approached from above by a B2 sector far field wave and from below by a C sector far field wave the limiting paths in $w_v x$ space form a closed region in S in which freely propagating waves coming in from, or going out to the far field cannot exist. This corresponds to a trapped domain. A study of flow over an asymmetric gully shows the existence of both a trapped region and a reflection region with reflections from C to B2 occurring on the turning point locus.

For the $\Sigma=-1$ wave group it may be shown that four critical frequancies occur: w_N, w_{BC}, w_P and w_X which are important for delineating the paths defining trapping and reflection. Most important is the fact that the type of signature which is theoretically predicted will depend on whether the propagation in the straining region corresponds to trapped or free domains in $w_v x$ space. As the S-region is crossed, in certain circumstances it is possible to encounter, for example, trapped B2 then C trapped then B2 trapped etc. waves. We then find that instead of the simple A (or B1) sector type signatures with which we are familiar (i.e. smooth sine wave-like forms) we obtain subsidiary maxima and minima or inflection points within the overall sine-like shape. The existence of this fine structure is particularly apparent for low wind and/or high strain rate situations.

For freely propagating C sector groups (i.e. Bragg frequency $w_S < w_N$) it is possible to obtain signatures which appear as dark bands alone (slicks) for suitable internal waves. Such sequences of dark bands (rather than alternating light and dark bands) are sometimes observed using SAR imagery. Such free C signatures occur for a relatively long wavelength radar (e.g. L-band) in low winds.

For V=1m/s and U_a=0.5m/s we obtain the following values for the critical frequencies

	w_N	w_{BC}	w_P	w_X
rad/sec	3	5	10	17
wavelength	7m	2.5m	63cm	22cm

For V=0.5m/s and U_a=0.25m/s the corresponding values are approximately

	w_N	w_{BC}	w_P	w_X
rad/sec	6	10	20	34
wavelength	1.75m	0.6m	11cm	5.5cm

Structure may therefore be expected for L- and X-band radars under the appropriate circumstances. If the radar backscatter also depends on wavelengths greater than the Bragg wavelength then complex signatures are even more likely.

Approaches which do not properly include refraction effects in low wind speed signature construction are not just wrong: they are totally wrong. Statements made in contemporary works that ignore wave refraction should be treated with extreme caution especially in the case of low wind speeds.

The case for $\gamma=0°$ was considered heretofore. For $\gamma\neq0°$ the situation is more complicated and it is not possible to give simple algebraic expressions for the important critical and other frequencies appropriate to the quadratic $\gamma=0°$ case. Instead a full treatment of the kinematic quartic is required and this is beyond the scope of the present outline. However the extra complications can be treated to advantage in a qualitative manner by reference to the $\Omega_v w$ space. This shows that for (almost) any current profile for $\gamma\neq0°$, both trapping and reflection regions exist in $\Omega_v w$ space at the same time.

Examination of computed signatures for various current types and various ocean environment parameters for both $\gamma=0°$ and $\gamma\neq0°$ cases reveals details of when and how the fine structure appears and how the bright bands (relative saturation $b>1$) diminish to leave only dark banded structure.

Some conclusions relating to the design or methodology of future experiments may be drawn from this work.

1. Sector A signatures are structure-free (i.e. smoothly continuous) and it is sector A with which we are concerned when the wind direction is the opposite to that in which the internal waves are travelling. Thus for studying the radar backscatter mechanism or the hydrodynamic interaction mechanism (e.g. wave-wave interactions) it is best to choose, where possible, times when the wind opposed the internal wave current direction. The opposite situation leads to complex signatures which would be far more difficult to interpret in terms of the above mechanisms.

2. Changes in the Phillips wave breaking exponent, n, produce very different signatures. It is suggested therefore that this parameter might be determined experimentally by comparing theoretical and observed signature shapes.

3. Changes in the shape of the equilibrium wave spectrum do not appear to be critically important in producing signature variations.

4. At low wind speeds, wave trapping via refraction effects can be very significant and can lead to complex signatures.

5. A reduction in the Phillips wave breaking exponent, n, is particularly effective in producing deep, wide slicks with only minimal narrow bright bands following the first bright band.

6. For an effective current of the Benjamin-Ono type extensive trapping occurs. Also the peaks of the free wave groups ($\Sigma = +1$) are noticeably separated from the peaks of the trapped, ($\Sigma = -1$) groups.

7. High wind speeds tend to produce decreased rip to slick modulations as does increasing the angle, γ, between the current and the range direction.

Acknowledgements: We thank Dr J.C. Scott and colleagues, ARE, Portland for their advice and encouragement in this research which was supported by the Procurement Executive, MOD.

Short wave modulation and breaking, experimental results

Siegfried Stolte

Forschungsanstalt der Bundeswehr
für Wasserschall- und Geophysik
Klausdorfer Weg 2-24
2300 Kiel 14
West Germany

Abstract

The small scale roughness of the sea is an important link in the energy flux from the atmosphere to the ocean. It determines growth and limitation of sea waves and its amplitude is modulated by wind, long waves, currents and instabilities.

To study these modulations experiments have been performed on board of the research platform NORDSEE. Wind, long and short waves, breakers and surface currents were measured. The analysis shows: Wave generation starts at a wind speed of 1.7 m/s at frequencies around 5 Hz and energy increases with wind up to 3.4 m/s. Short waves are modulated on long waves and spectral decays are near $f^{-4.5}$, and f^{-2} at frequencies above 8 Hz.

The modulation strength grows with wind speed and wave height and is a function of frequency with a minimum near 8 Hz. Positions of energy maxima are widely scattered on long waves. Second maxima are often developed.

Breaker activities and intensities increase with wind and long wave slope, while the frequency of breaking decreases with wave length.

Short wave modulation and breaking can differ under similar conditions. They cannot be explained by wind or wave height alone. Interpretations have to include the parameters of the interaction system atmosphere-ocean, particularly wave slope and atmospheric stability criteria.

1 Introduction

A turbulent wind transfers energy to a smooth sea surface and excites ripples in the cm- and dm-regime. Their amplitudes are in equilibrium shortly while longer components grow and further modulation processes become effective. Under moderate wind conditions modulation is caused by wind, nonlinear interaction, and micro-breaking as consequence of intensified drift. High amplitudes are expected in front of crests and on windward sides.

At high wind speeds crests become steep and asymmetric. Capillary waves are generated in front of them, micro-breaking is intensified. When long waves get unstable and break, a jet forms and short scale roughness and bubbles are generated. Turbulent patches remain in which ripples are suppressed while long wave phase proceeds. After a while turbulence dissipates and short waves gain higher amplitudes.

G. J. Komen and W. A. Oost (eds.), Radar Scattering from Modulated Wind Waves, 201–210.

Significant modulation parameters are:

o wind
o nonlinear interaction
o micro-, wave breaking
o turbulence, rain, surface films, bubbles

Each of these components can be effective, specific experiments have to be conducted.

2 Measurements

Two series of measurements were performed aboard of the research platform NORDSEE, a modulation and a breaker experiment. According to requirements and available instruments measuring setups had been chosen in the following way:

o two cup anemometers, 10 m and 4 m above sea level
o an additional ultrasonic anemometer in 5 m elevation during the breaker experiment (three components and temperature)
o a resistence-wire measuring device for low- and high-frequency wind waves (5.5 m length)
o a current meter 3.5 m below sea level; during an additional experiment as wave follower (5 - 10 cm below surface, .1 s time constant).

The instruments were installed in a frame and launched by a crane 20 m off the platform, stabilized by a weight of 600 kp 17 m below sea level. The orientation with respect to wind and current was always chosen in such a way that no disturbances, caused by platform or frame, could be observed.

3 Data evaluation and results

3.1 Modulation analysis

It was the aim to study short wave energy variation depending on wind and phases of long waves.Data were analysed as follows: approximation of long waves by sine waves and computation of short wave energy spectra and wind parameters as a function of phases of long waves.

Data were measured in a fixed position. Therefore high-frequency spectra in long wave facets are frequency-shifted in a nonlinear way:

$$n = \sigma(\vec{k}) + \vec{k} \cdot \vec{v}$$

n = measured frequency, frequency of encounter
$\sigma(\vec{k})$ = intrinsic frequency at wavenumber $\vec{k}$
$\vec{v}$ = current velocity

The current consists of three components:

$$\vec{v} = \vec{v}_{u.o.} + \vec{v}_{orb.} + \vec{v}_{dr.}$$

$\vec{v}_{u.o.}$ = current of the uppermost ocean
$\vec{v}_{orb.}$ = orbital current of long waves
$\vec{v}_{dr.}$ = wind-induced drift current in top mm-layer

The orbital current in facet positions is calculated from the measurements in 3.5 m depth, the drift velocity as 3% of the instantaneous 10 m level wind (Evans and Shemdin, 1980). Short waves and wind are assumed to be in the same direction so that only the current component in wind direction is relevant. All data are corrected for phase shifts and time delays induced by the electronic filters and recording instruments.

Intrinsic frequency spectra in long wave positions are computed as follows: long waves of 2 s period and more are approximated. At phase angles of $0^o, 15^o$ etc. currents are calculated, facet slopes and accelerations from the approximations. Spectra of encounter $E(n) \cdot \Delta n$ are computed in the frequency range from 2 to 32 Hz (32 values, .5 s time series). They are converted into intrinsic spectra by the equations:

$$k^3 - \frac{v^2}{\gamma} k^3 + \frac{\tilde{g} \cdot 2 \cdot n \cdot v}{\gamma} \cdot k - \frac{n^2}{\gamma} = 0$$

and

$$E(\sigma) = E(n) \frac{\Delta n}{\Delta \sigma}$$

$\tilde{g}$ = acceleration perpendicular to the facet
$\gamma = \frac{\tau}{\rho}$ = surface tension / water density

The first equation can be solved in case of positive currents and negative ones up to about .15m/s. For analysis the time series were chosen in such a way that wind and current were in the same direction.

In addition friction velocities were calculated in phase angles from the measurements in two heights by using the equation of neutral wind profile.

Analyses were performed from 24 min time series.

Results

Conversion of spectra results in characteristic modifications: Intrinsic spectra show less energy than those of encounter and frequency decays are wind-independent. In the range up to 8 Hz they are shifted to a $f^{-4.5}$-decay, at higher frequencies they follow a f^{-2}-characteristic. These different saturation ranges cannot be seen in spectra of encounter, the frequency decays are nearly uniform and wind dependent (fig. 1).

Computations of high frequency spectra without correction of Doppler-shift result in most cases in distorted spectra. Currents cause nonlinear energy transformations which cannot be balanced by averaging, not even in pure sine currents.

An example of short wave development is shown in figure 2: The wind decreases and after calm it blows up to 4 m/s. Intrinsic spectra at crest position show the swell situation, excitation of waves at a threshold of 1.7 m/s wind speed, energy increase up to 3.4 m/s and decrease though the wind grows to 4m/s. Now a spectral dip is seen around 11 Hz.

The reduction of energy coincides with the onset of breaking.

Computations of phase dependent spectra result in modulations represented by an example (fig. 3). Spectra of encounter and intrinsic spectra are shown at long wave crest (0°) and trough position (180^0). In this situation short waves are more energetic at the trough than at the crest, while uncorrected spectra always have their maxima near crests.

The ratio of energy densities at frequencies and phase angles $E(f,\phi)$ to the mean E(f) describes phase dependent modulations (fig. 4). First modulations of uncorrected frequencies from 2 to 8Hz are given, which show maxima near crests - a little to the windward side - and minima at troughs.

Modulations of intrinsic frequencies show that the maxima are shifted to 240° on the windward side.

Similar results were found at frequencies from 10 to 16 Hz, in some cases second maxima are developed in front of the crests.

Summarily, the characteristics of short wave modulation are:

Positions of energy maxima are largely scattered but predominantly situated on windward sides of long waves, minima are 180° phase shifted. The scatter increases with frequency and maxima are found more often in front of the crests, especially at 14 and 16 Hz. Second maxima are often developed. They are located on both crest sides.

A relation between phase angles of extrema and wind or wave height cannot be deduced from these results.

Defining the modulation strength by the ratio of maximum to minimum energy, relations with respect to wind and wave height exist (fig. 5): The modulation strength grows with wind and waves. It is frequency dependent with lowest increasing rates around 8 Hz where the transition from $f^{-4.5}$ - to f^{-2} - decay occurs.

The qualitative comparison of phase dependent friction velocity and short wave modulation results in the following observations: At moderate wind speed short waves and friction velocity are modulated in the same way, i.e. in high energy regions on windward sides friction velocity increases too, and vice versa. When wave breaking occurs maxima of friction velocity are in front of crests and coincide more closely with minima of wave energy.

The conclusion from this analysis is that micro - and wave breaking are most important processes in short scale roughness modulation. Breakers have to be analysed in more detail.

3.2 Breaker analysis

Waves get unstable and break when the particle velocity near the crest exceeds that of the phase speed. Water flows down on foreward sides. Current velocities can exceed 10 m/s when long waves break. These events are visible by foam production and breakers were marked on magnetic tape during the experiment.

A breaker threshold was determined by means of a differentiator. More precise computer optimizations resulted in thresholds of the begin and end of breakers so that well defined time series were available for analysis.

In order to calculate breaker momentum water mass and velocity must be known.

The signals represent primarily time dependent surface elevations and, after application of threshold criteria, time intervalls of breakers.

The momentum is given by:

$$m_{br} = \rho_{br} \cdot V_{br} \cdot u_{br}$$

$\rho_{br}, V_{br}, u_{br}$ = density, volume, velocity of the breaker

Breaker velocities can be estimated from the condition:

$$u_{br} > c_w$$

c_w = phase speed of breaking wave

The amount of air in the turbulent jet is assumed to be less than 30% . The velocity exceeds that of the wave speed not too far:

$$\rho_{br} \cdot u_{br} \approx \rho_w \cdot c_w$$

ρ_w = density of water

Similarly the volume per unit width is approximated by:

$$V_{br} \approx \frac{h_{br} \cdot l_{br}}{2}$$

$$l_{br} = u_{br} \cdot t_{br} \approx c_w \cdot t_{br}$$

h_{br}, l_{br}, t_{br} = height, length, time of the breaker

and the momentum:

$$m_{br} \approx \rho_w \cdot c_w \cdot \frac{h_{br} \cdot l_{br}}{2}$$

To calculate u_{br} from the wave approximation it has to be assured, that breakers observed and defined by the thresholds are those of the approximated waves and not short components which break at crests of carrier waves. Comparative measurements have been carried out with a wave follower which gave agreement of estimated and measured velocities (fig. 6).

Results

Momentum distributions have been calculated from 21 min time series: Breaker activities decrease with wave length and momentum distributions are limited by an upper boundary (fig. 7). Moreover frequencies of breaking and momentum increase with wind and long wave slope.

Considering the energy balance of the wave field it has to be mentioned that breaker and orbital current are in the same direction. Only part of the breaker momentum is dissipated, the rest enters the wave field again.

4 Conclusions

Short wave modulation and breaker activity cannot be explained by wind or wave height alone. Interpretations have to include more parameters of the system atmosphere-ocean, e.g. wave slope and atmospheric stability.

Waves get steeper when they "feel" the bottom or run into countercurrents. Modulation and breaking are intensified, and vice versa. Examples of radar pictures show these modulations.

Reference

Evans, D.D. and Shemdin, O.H. 1980,
An Investigation of the Modulation of Capillary and Short Gravity Waves in the Open Ocean.
J. Geophys, Res. 85

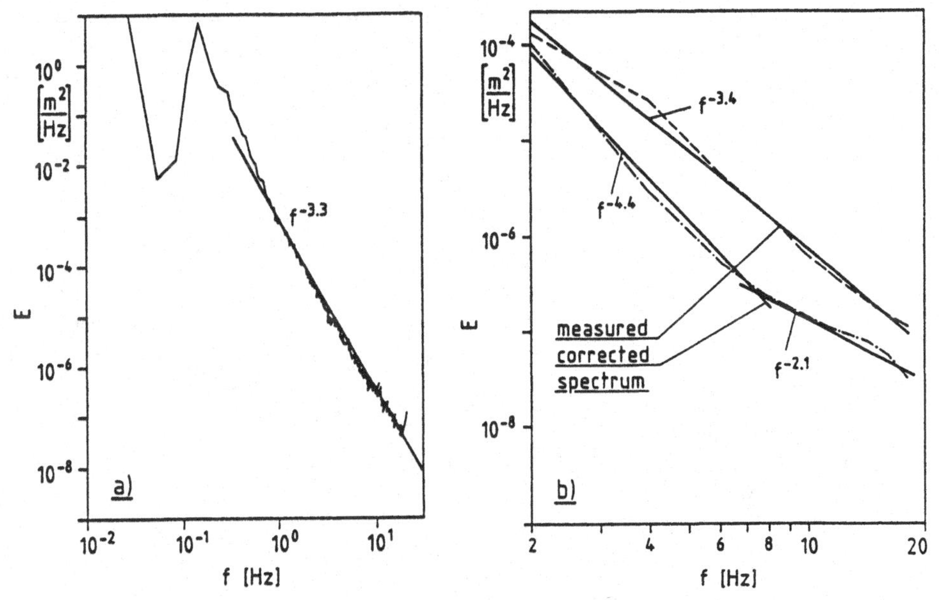

Fig. 1: Spectrum of encounter (a) and Doppler-correction of high frequency spectra at crest position of long waves (b) (wind:19 m/s, wave height:1.6 m)

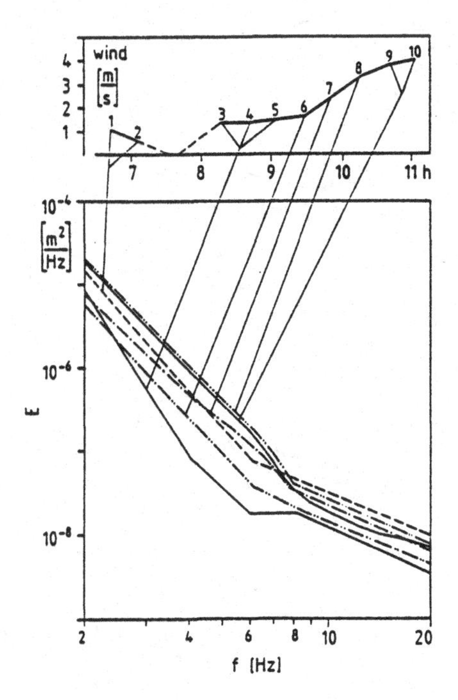

Fig. 2: Short wave development

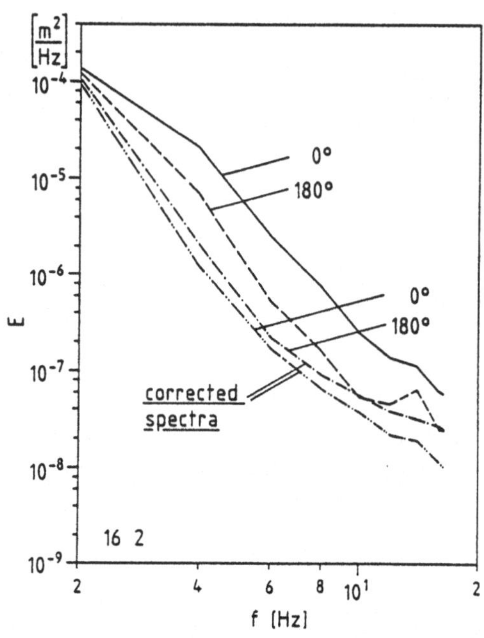

Fig. 3: Spectra of encounter and intrinsic spectra at crest (0°) and trough position (180°) (wind:10 m/s, wave height:1 m)

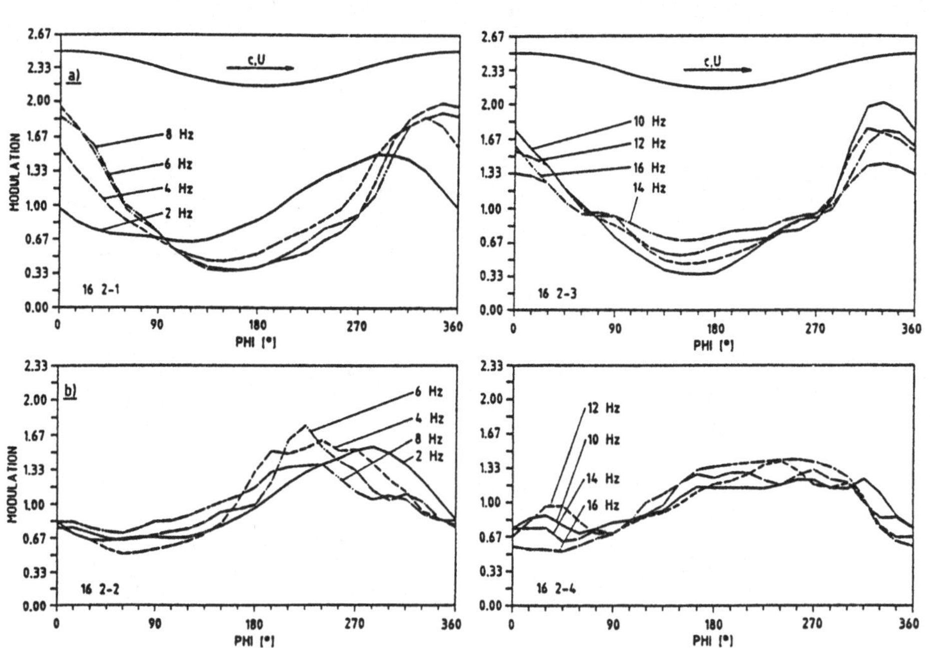

Fig. 4: Phase dependent short wave modulation: Doppler-shifted (a) and corrected functions (b) (same measurement as fig. 3)

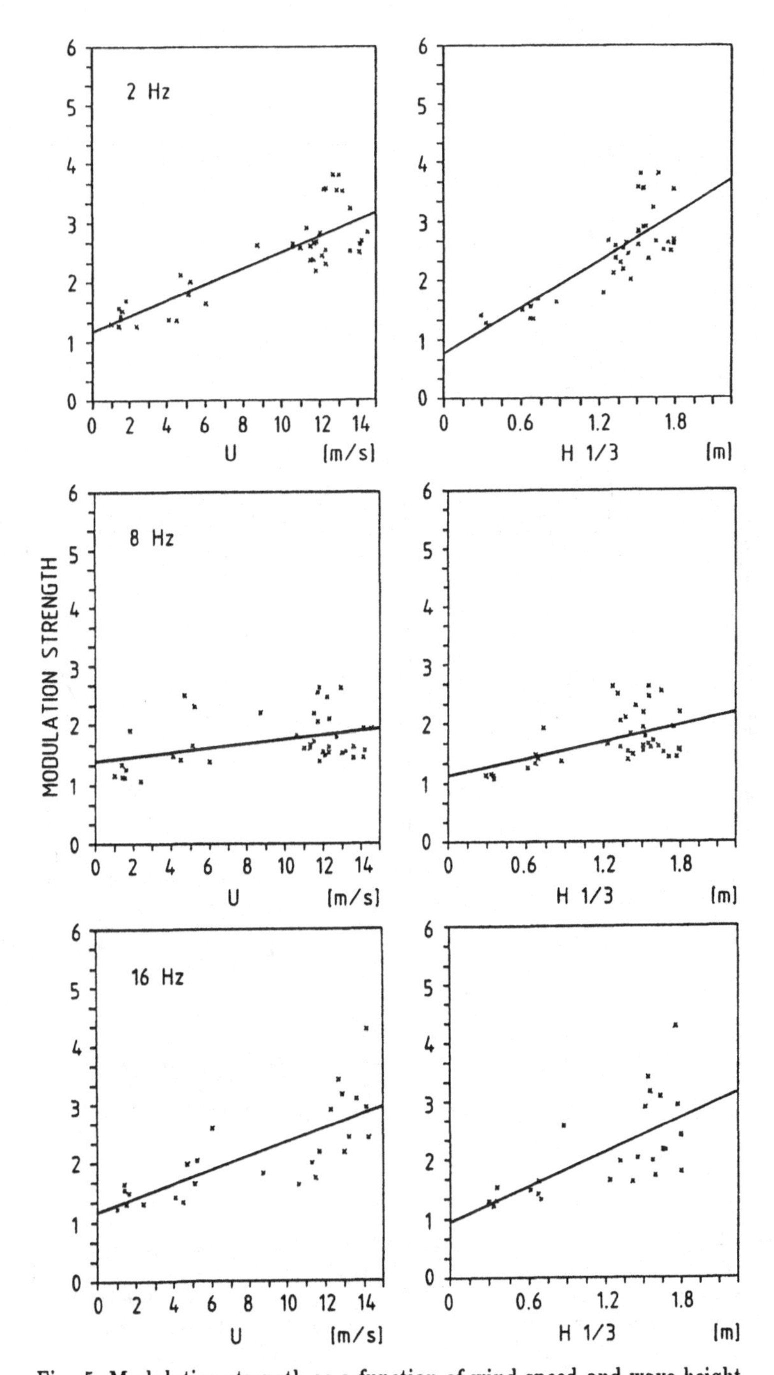

Fig. 5: Modulation strength as a function of wind speed and wave height

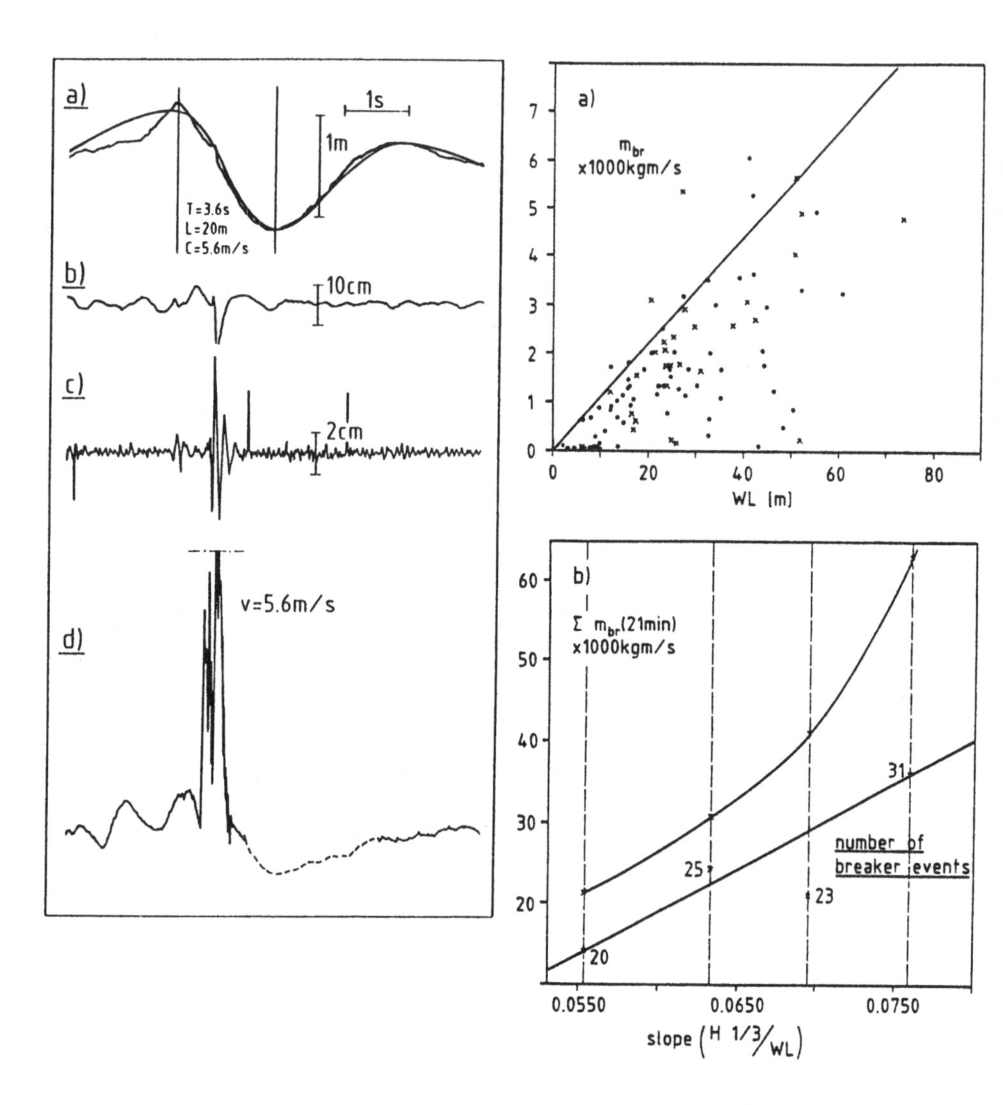

Fig. 6: Example of breaker measurement
a) long wave signal
b) 1 Hz high pass filtered wave signal
c) 5 Hz high pass filtered wave signal
d) surface current velocity

Fig. 7: a) Distribution of breaker momentum (84 min time series, 1 m unit width)
b) Total breaker momentum of 21 min time series related to long wave slope (same data set as (a))

Nonlinearity as the missing link between wavelets and currents

Klaartje van Gastel
Mathematisch Instituut RUU

Postbus 80010 3508 TA Utrecht
the Netherlands

Abstract

To describe the large modulation of short surface waves by nonuniform currents as encountered for weak winds a two timescale technique is applied to the energy balance for short waves. A scaling is assumed in which wind input, nonlinear interactions, dissipation and horizontal current shear are all of the same strength. This approach leads to modulations correct in order of magnitude.

1 Introduction

Short surface waves, with wavelengths of less than about 20 cm, are closely linked to the surface current. This is made apparent by radar images of the sea (De Loor 1981, Fu & Holt 1982). The strength of this coupling, as expressed by the ratio of the modulation of the energy density of the surface waves to the modulation of the current, is far from being constant: it ranges from about 1 (Alpers et al. 1982, Kasischke 1986) to 100 or 1000 (Valenzuela et al. 1985, Halsema et al. 1986, Kwoh et al. 1986). These large coupling coefficients have been encountered for short waves, X and K band, and low windspeeds (less than 5 m/s). In this paper the two timescale method is used to derive equations describing these large modulations.

G. J. Komen and W. A. Oost (eds.), Radar Scattering from Modulated Wind Waves, 211–218.

2 Scaling of the equations

The development of the wave spectrum is commonly described by the energy balance (Hasselmann 1960, Willebrand, 1975):

$$[\frac{\partial}{\partial t}+\frac{\partial\Omega}{\partial k}\frac{\partial}{\partial x}-\frac{\partial\Omega}{\partial x}(\frac{\partial}{\partial k}+\frac{1}{k}(\frac{\partial\omega}{\partial k}\frac{k}{\omega}-1))]G=S_{IN}+S_{NL}+S_{DIS} \quad (1)$$

$$\Omega=kU+\omega \quad (2)$$

Here G is the variance density, related to the energy density E by $G=(2\pi k/\rho\omega^2)E$, k is the wavenumber, ω the intrinsic frequency and U is the surface current. The three source terms in the right hand side still have to be specified, they stand resp. for the combined effect of wind and viscosity, nonlinear interactions between the waves and for dissipation. In this paper equation (1) will only be considered under the assumptions of one-dimensionality and stationarity in time.

To further simplify the energy balance, it is common to assume linearity (Hughes 1978, Lyzenga et al. 1983, Alpers & Hennings 1984, Yuen et al. 1986, Ermakov & Pelinovsky 1984). Physically, this assumption is equivalent to assuming 1. that the sytem starts near an equilibrium and 2. that it remains near this same equilibrium. Therefore, this assumption is applicable when the modulation is order 1, but it is evidently not justified when the modulation is order 100.

An approach which does fit in with large modulations is to assume a slowly varying quasi-equilibrium. The slow variations are generated by changes in the current, which changes indeed can be thought of to be slow compared to the period of the waves (Alpers et al. 1981). These slow variations give rise to small terms in the equations. Therefore, to keep the approximation scheme consistent, one also has to take account of other small terms, e.g. terms quadratic in the steepness of the waves or proportional to ρ_a/ρ_w.

This approach is commonly indicated by the name of two time-scale technique (Eckhaus 1979). It can be formalized as follows: Equation (1) is expanded according to

$$\left.\begin{array}{l} G = \epsilon_1 G_1(x_0, x_1, k) + \epsilon_1^2 G_2(x_0, x_1, k) + \ldots \\ U = U_0(x_1) \\ \beta = \epsilon_2 \beta_1 \\ \frac{\partial}{\partial x} = \frac{\partial}{\partial x_0} + \epsilon_3 \frac{\partial}{\partial x_1} + \ldots \end{array}\right\} \quad (3)$$

Note that S_{NL} is quadratic in G (see below), this is indicated by S_{NL}(G,G). Of the function S_{DIS} little is known (Phillips 1984); I will assume that

$$S_{DIS} = \epsilon_1^2 S_{DIS2} \quad (4)$$

It is seen that three small parameters occur in the equations: the steepness ϵ_1, the ratio of wind induced growth to frequency ϵ_2, and the ratio of wavelength to length scale of the current variation ϵ_3. For $u_* = 0.25 m/s$ or less, low sea state and a length scale of the current variation of 10 to 100 m I expect these three will typically be of the same order. Note that these are the circumstances under which the large modulations have been encountered.

This situation will be considered in the rest of this paper, thus

$$\epsilon_1 \simeq \epsilon_2 \simeq \epsilon_3 \quad (5)$$

Inserting (3)/(5) in (1), the following equations are obtained:

$$1^{st}\text{order}: \ (U_0 + c_g)\frac{\partial}{\partial x_0} G_1 = 0 \quad (6)$$

$$2^{nd}\text{order}: \ (U_0 + c_g)\frac{\partial}{\partial x_0} G_2 = -(U_0 + c_g)\frac{\partial}{\partial x_1} G_1$$

$$+\frac{\partial U_0}{\partial x_1}(k\frac{\partial}{\partial k} + \frac{k c_g}{\omega} - 1)G_1 + \beta_1 G_1 + S_{NL}(G_1, G_1) + S_{DIS2} \quad (7)$$

To avoid secular behaviour for G_2, an extra equation results:

$$(U_0 + c_g)\frac{\partial}{\partial x_1} G_1 - \frac{\partial U_0}{\partial x_1}(k\frac{\partial}{\partial k} + \frac{k c_g}{\omega} - 1)G_1 =$$

$$= \beta_1 G_1 + S_{NL}(G_1, G_1) + S_{DIS2} \quad (8)$$

Equation (6) shows that on the fast space scale there is equilibrium, while equation (8) determines the development of this equilibrium on a slow space scale.

3 Results and interpretation

For S_{IN} and S_{NL} the following expressions can be derived (Gastel et al. 1985, Gastel 1987A):

$$S_{IN} = \beta(u_*, U, k)G \tag{9}$$

$$S_{NL}(k) = \int\int J\delta(k - k' - k'')\delta(\omega - \omega' - \omega'')dk'dk'' \tag{10}$$

Here J depends on k, k', k'' and the variance levels at these wavenumbers. S_{DIS} I have parametrized by

$$S_{DIS} = \begin{cases} -S_{IN} & S_{IN} > 0 \\ 0 & S_{IN} < 0 \end{cases} \tag{11}$$

This parametrization has been chosen for its convenience. Luckily, the results are insensitive to this choice (Gastel 1987 B).

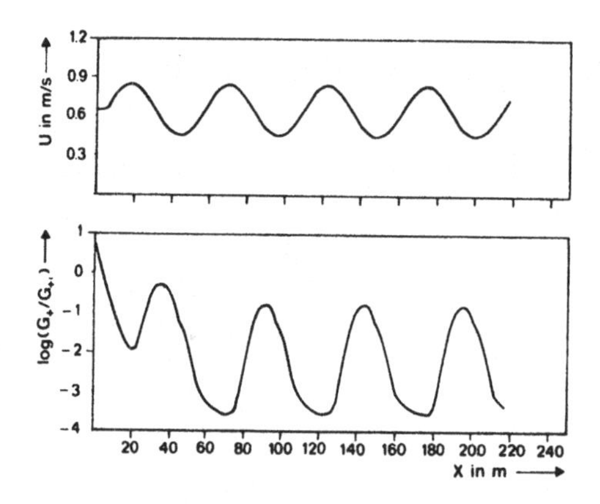

Figure 1. For a current field given by $U = 0.66 + 0.2sin(2\pi x/52)$, SI units, the modulation of the variance level of short waves is calculated. Here the results are shown for $k = 310\ m^{-1}$. (from Gastel 1987B)

Using the model determined by (2),(8)/(11) space integrations have been performed numerically. Several current fields have been taken as input. In all cases, the outcome was small modulation for $k \leq 260\ m^{-1}$ and large modulation for $k \geq 260\ m^{-1}$ (Gastel 1987B). A typical run is shown in figure 1. Qualitatively, there is agreement with data in phase and depth of the modulation (Gastel 1987B). This can be considered as a check on the usefulness of assumption 5.

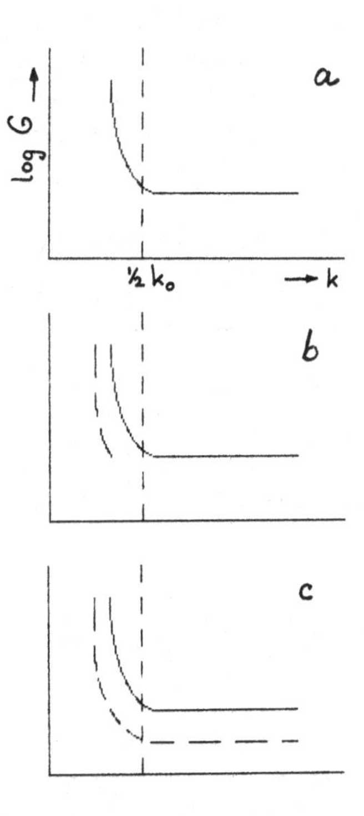

Figure 2. A cartoon of the feedback induced by the combination of non-linear interactions and horizontal current shear. a: The nonlinear interactions drive the spectrum to an equilibrium shape with a sharp bend at $k = \frac{1}{2}k_0$. b: The current gradient shifts the spectrum. c: The nonlinear interactions reinforce the equilibrium shape.

The very large modulation factors seen in figure 1 are somewhat startling. One would like to understand on intuitive grounds why equation (8) leads to this behaviour. The following considerations might give some insight. The nonlinear actions tend to drive the spectrum very strongly to an equilibrium shape (not height), as scetched in figure 2a. A sharp bend is induced at $\frac{1}{2}k_0 = \sqrt{(g/2T)} \simeq 260\ m^{-1}$, g being the gravity force and T the ratio of surface tension to density. This phenomenon is related to k_0 being a bifurcation point for the nonlinear interactions (Gastel 1987A). The direct effect of the horizontal current shear is to produce a shift on this spectrum, to the left for $U_x > 0$. This can be seen by solving eq. (8) for zero right hand side. This shift drastically changes the energy level just to the left of $\frac{1}{2}k_0$ (figure 2b). The nonlinear interactions will then reinforce the equilibrium shape, implying a strong drop in energy level for all $k \geq \frac{1}{2}k_0$ (figure 2c).

Summarizing, the combined effects of nonlinear interaction and horizontal current shear leads to feedback in tne wave system, which makes large modulations possible. Phenomenologically, this feedback can be described by

$$\frac{\partial}{\partial x}G = c(\frac{\partial}{\partial x}U)G \tag{12}$$

where c is some constant. For this feedback to occur, the nonlinear interactions are crucial.

4 Conclusions

Data of modulation of the energy of short surface waves by nonuniform currents show very different behaviour for weak winds (< 5 m/s) versus moderate winds (5-10 m/s). Therefore, one would expect different scaling to apply to the equations in these two regions. Indeed, for moderate winds linear models of, for instance, Alpers & Hennings (1984) or Hughes (1978) can give agreement with data, while for weak winds a nonlinear model as discussed here can lead to good results. This supports the assumption that for weak winds, low sea state and length scales of the current variation of 10 to 100 m the effects of wind input, nonlinear interactions, dissipation and horizontal current shear are all of the same order.

References

Alpers, W., and I. Hennings, *A theory of the imaging mechanism of underwater bottom topography by real and synthetic aperture radar, J. Geophys. Res 89(C6),10,529-10,546,1984.*

Alpers, W., D.B. Ross, and C.L. Rufenach, *On the detectability of ocean surface waves by real and synthetic aperature radar, J. Geophys. Res., 87(C7),6481-6498,1981.*

Eckhaus, W., *Asymptotic analysis of singular perturbations, North-Holland, Amsterdam 1979.*

Ermakov, S.A., and E.N. Pelinovsky, *Variations of the spectrum of wind ripples on coastal waters under the action of internal waves, Dyn. Atmos. Oceans, 8,95-100,1984.*

Gastel K. van, *Nonlinear interactions of gravity-capillary waves: Lagrangian theory and effects on the spectrum, J. Fluid Mech., 182,449-523,1987A.*

Gastel, K. van, *Imaging by X band radar of subsurface features: a nonlinear phenomenon, J. Geophys. Res. 92(C11), 11,857-11,865,1987B.*

Gastel K. van, P.A.E.M. Janssen, and G.J. Komen, *On phase velocity and growth rate of wind-induced gravity-capillary waves, J. Fluid Mech., 161,199-216,1985.*

Halsema D., van, A.L. Gray, S.J. Hughes and B.A. Hughes, *C- and Ku-band scatterometer results from the scattmod internal wave experiment, Eur. Space Agency Spec. Publ., ESA SP-254,311-317, 1986.*

Hasselman, K., *Grundgleichungen der Seegangsvorhersage, Schiffstechnik,7,191-195,1960.*

Hughes, B.A., *The effect of internal waves on surface wind waves, 2, Theoretical analysis, J. Geophys. Res.,83(C1),455- 465,1978.*

Kasischke, E.S., *Characterisation of internal wave surface patterns on airborne SAR imagery, Eur. Space Agency Spec. Publ., ESA SP-254,1055-1060,1986.*

Kwoh, D.S.W., B.M. Lake and H. Rungaldier, *Identification of the contribution of Bragg-scattering and specular reflection to X-band microwave backscattering in an ocean experiment, Space and Technol. Group, TRW, Redondo Beach, Calif. 1986.*

Lyzenga, D.R., R.A. Shuchman, E.S. Kasischke and G.A. Meadows, *Modeling of bottom-related surface patterns imaged by synthetic aperture radar, paper presented at International Geoscience and Remote Sensing Symposium '83, IEEE, San Francisco, Calif., 1983.*

Philips, O.M., *On the response of short ocean wave components at fixed wavenumber to ocean current variations, J. Phys. Oceanogr., 14,1425-1433,1984.*

Valenzuela, G.R., W.J. Plant, D.L. Schuler, D.T. Chen and W.C. Keller *Microwave probing of shallow water bottom topography in the Nantucket Shoals, J. Geophys. Res. 90(C3),4932- 4942,1985.*

Willebrand, J., *Energy transport in a nonlinear and inhomogeneous random gravity wave field, J. Fluid Mech., 70,113- 126,1975.*

Yuen, H.C., D.R. Crawford and P.G. Saffman, *SAR image of bottom topography in the ocean: Results from an improved model, Eur. Space Agency Spec. Publ., ESA SP-254,807-812,1986.*

THE TRICKLE DOWN OF WAVE MODULATIONS FROM SWELL TO GRAVITY CAPILLARY WAVES

David Sheres
Naval Research Laboratory
Washington, DC 20375-5000

ABSTRACT. Modulation of the wavenumber and amplitude of swell by mesoscale current and bathymetry features leads in turn to patchiness of the short gravity capillary wave field due to the modulated swell. A selective review of these processes and their implication to air sea interaction and remote sensing is presented.

1. BACKGROUND

Short gravity-capillary (G-C) waves are spatially modulated by a number of factors such as non uniform currents (e.g. surface and internal gravity waves), spatial variability of the wind, stability of the boundary layer above, surface films, dissipation, turbulence and nonlinear interactions between the waves. The resulting patchiness of the short wave field affects the remote sensing of the ocean by most sensors; sometimes it appears as noise, in radiometers for example (Hollinger, 1971). Sometimes it is the desired signal reflecting the ocean processes that we want to measure, as the wealth of features observed by SEASAT SAR and Scully-Powers photographs of the ocean from the shuttle (Scully-Powers, 1985) gives evidence. In order to understand the nature of both the signal and the noise aspects of the short wave patchiness, we have to investigate the complex interactions creating it.

In this paper we review selectively the cascade of events from mesoscale surface current features and bathymetry to much smaller scale patchiness in the short wave field, including its effects on wind forcing and potential feedback mechanisms between surface flow, surface waves and wind stress. We can not, I believe, yet quantify every step along this cascade, but we do have a conceptual understanding of the processes.

The first part of this paper describes the modulation of swell by mesoscale current features and bathymetry. Next, the modulation of the short G-C waves by swell in the simple case of no wind forcing or dissipation is described. The last part briefly addresses the

G. J. Komen and W. A. Oost (eds.), Radar Scattering from Modulated Wind Waves, 219–229.

implications and applications of the resulting patchiness. This review is selective and reflects the bias of my familiarity and involvement.

2. THE INTERACTION OF SWELL WITH MESOSCALE CURRENT AND BATHYMETRY FEATURES

Swell is refracted by mesoscale surface current features, such as Gulf Stream undulations and eddies, with scales on the order of 100 km. Swell is also refracted by shallow bathymetry features, such as the Nantucket Shoals (Valenzuela et al., 1985) which have horizontal dimensions of about 20 km and depth of less than half a wavelength of the swell. The refraction patterns exhibit characteristic regions of increase in wave amplitude (wave ray focusing) adjacent to regions of reduced wave amplitude as well as changes in the swell's wavenumber. The dimensions of these focusing areas is calculated to be on the order of a few km, much smaller than the refracting features (Sheres et al., 1987).

The calculation of these refraction patterns follows some form of the ray equations, with the underlying assumptions of ray optics that the waves are almost plane waves, or that wave parameters change very little over a distance of one wavelength. Kenyon (1971) first applied these equations to calculate swell refraction by the equatorial and circumpolar currents. Teague (1974) used Kenyon's formulation to calculate wave refraction by an eddy. Dobson and Irvine (1983), Mathiessen (1987), Sheres et al. (1987) and Sheres and Kenyon (1988) addressed various aspects of the refraction of swell by eddies.

In what follows, I will show numerical results of swell refraction by mesoscale flow features. The ray equations are:

$$\sigma = \sigma_o + \vec{k} \cdot \vec{U} \tag{1}$$

$$C_{g_i} = \frac{dx_i}{dt} = \frac{\partial \sigma}{\partial k_i} \tag{2}$$

$$\frac{dk_i}{dt} = - \frac{\partial \sigma}{\partial x_i} \quad \text{where} \quad \frac{d}{dt} = \frac{\partial}{\partial t} + C_{g_i} \frac{\partial}{\partial x_i} \tag{3}$$

where $\vec{k}$ is the wavenumber, $\vec{U}$ is the water velocity, $\vec{C}_g$ and σ are the absolute group velocity and frequency of the waves (measured from a fixed platform) respectively; a subscript "o" means that the quantities were measured from a platform moving with the flow velocity $\vec{U}$. Numerical integration of these ray equations results in ray refraction patterns for swell passing through the distribution of $\vec{U}$ chosen as an input for that calculation. The rays are the paths of wave energy propagation and are usually different from the wave orthogonals, which are perpendicular to the crests.

Figure 1 shows the outline of the Gulf stream and associated eddy field produced by NOAA from Infra-Red imagery of the ocean. Figure 2 shows the idealized velocity distribution of a Gulf Stream undulation and figure 3 shows the ray refraction pattern of 100m waves passing through this undulation, obtained by numerically integrating the ray equations. A region of ray convergence is evident at the lee of the flow feature suggesting a "hot spot" of wave energy there. This focus of wave energy (caustic) is typical for ray refraction patterns by undulating or rotating current features. Figure 4 shows the refraction details of 100m wave rays by an eddy with a linear velocity distribution varying from zero at the center to a maximum of 2.5 m/s at a radius of 30 km and back to zero at a radius of 50 km. The ray undulations near the radius of maximum velocity are due to the reversal in the shear direction encountered by the waves as they cross that radius. One wave ray that starts at x~125 km crosses the x axis again at x=32 km; thus some wave energy can be totally reflected by an eddy. The refraction patterns depend on the scales of the waves and the refracting velocity features. When the ratio of flow velocity to group velocity of the waves remain constant, so does the refraction pattern, provided the spatial scales of the flow remain constant as well (Sheres and Kenyon, 1988).

As the rays indicate wave energy paths, no energy propagates in a direction perpendicular to the rays. This suggests that the ray refraction patterns contain information about wave energy, provided there is no dissipation or generation (by wind or other interactions) of wave energy. The wave energy can thus be calculated via the Jacobian J, that is proportional to the area of a ray tube element (the product of $C_g dt$ and the distance to an adjacent nearby ray, Christoffersen, 1982).

$$\left(\frac{E_1}{E_2}\right)^{1/2} = \left(\frac{k_1}{k_2}\right)^{1/4} J^{-1/2} K_f \tag{4}$$

where J is a measure of ray refraction and can be calculated as

$$\frac{1}{J}\frac{dJ}{dt} = \nabla \cdot \vec{C}_g \tag{5}$$

and E is the wave energy density and K_f is the dissipation parameter that accounts for sources and sinks (here assumed $K_f=1$), C_g is obtained by integrating equation 2. The indices 1 and 2 refer to positions along the ray.

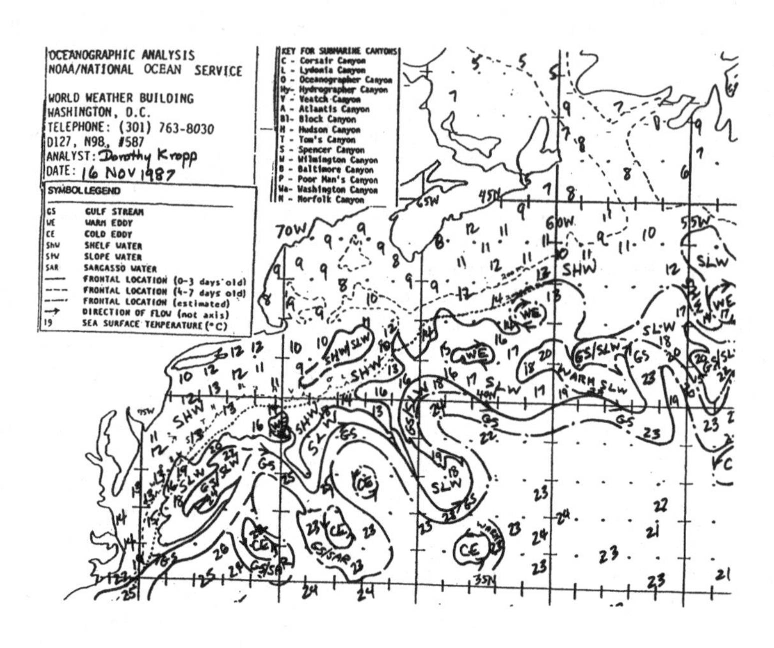

Figure 1. An outline of the Gulf Stream and associated eddy field produced by NOAA, mostly from infra red satellite imagery.(courtesy Jenifer Clark)

How does focusing of swell energy affect the gravity-capillary waves? The behaviour of swell in such focusing regions and beyond does not lend itself to simple or elegant mathematical solutions. It is clear however that long wave amplitude is enhanced in the focus region while at nearby regions wave amplitude is reduced, as compared to the initial waves before refraction. The ambient G-C waves encounter these modulations in swell as variability in surface velocity. A lot of work has been done on the evolution of G-C waves in the presence of a velocity distribution, such as presented by underlying swell, and by the input of wind, "output" by dissipation and non-linear interactions. (Recent work is presented by a number of papers in this book). I will briefly review here a fairly "benign" case of the refraction of G-C waves by swell, when there is no wind, negligible dissipation and no nonlinear interactions. This can give a picture of such two dimensional interaction and offer a comparison for the more sophisticated approaches based on the conservation of energy density (or wave action), when the so called "right hand side" is negligible.

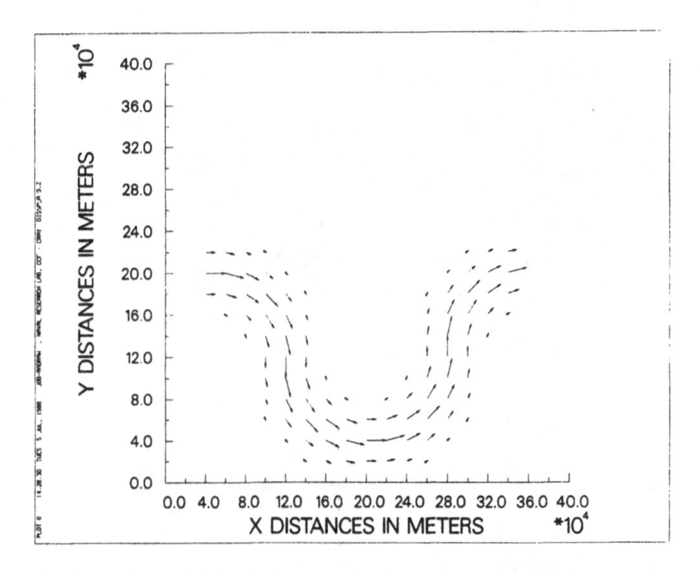

Figure 2. Idealized velocity distribution of a Gulf Stream undulation. The velocity varies linearly from zero at the edges to a maximum of .75 m/s at the center.

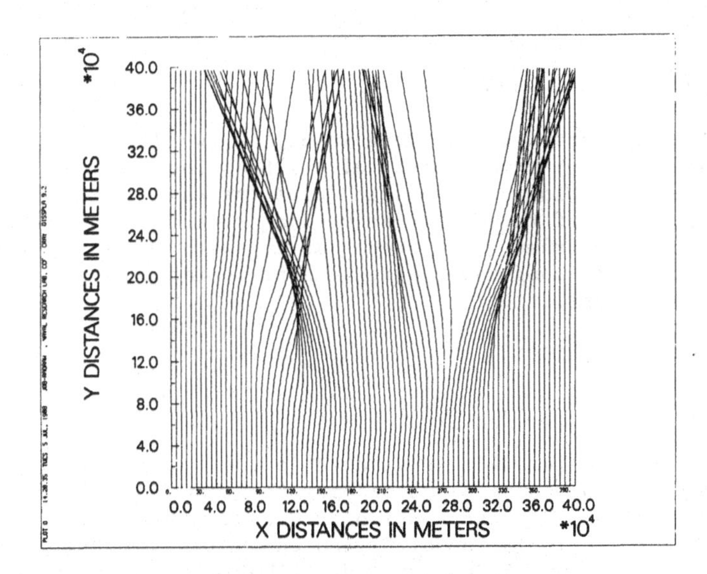

Figure 3. Refraction of 100 m wave rays by the undulation shown in figure 2. The rays are initially directed up, at the bottom of the figure, and refract as they propagate through the undulation.

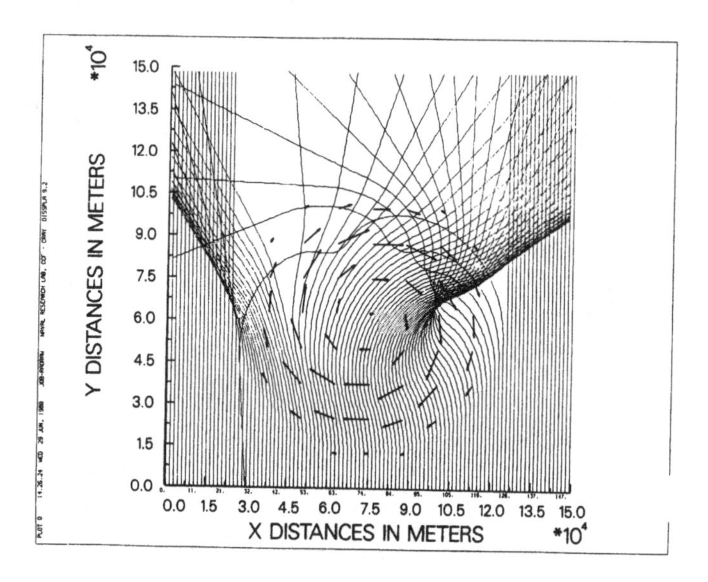

Figure 4. Details of ray refraction of 100 m waves by an anticyclonic eddy with a linear velocity distribution varying from zero at the center to 2.5 m/s at a radius of 30 km, and back to zero at a radius of 50 km. Vectors describing the velocity distribution in the eddy are superimposed on the rays.

3. SHORT WAVES ON LONG WAVES

The refraction of short waves by long waves, in two dimensions, was calculated analytically by Kenyon at al. (1983). They assumed no dissipation or generation of waves, and used the ray equations (1-3) to calculate refraction of the G-C waves by long waves. The calculation was carried out in a frame of reference moving with the long waves phase velocity C; the long waves were thus presented as a steady flow U with a velocity distribution:

$$U(x)=C(1-B\cos Kx) \tag{6}$$

where

$$B=gA/C^2$$

The gravitational acceleration is g and A is swell amplitude. When A is small and linear theory applies, B becomes the long wave slope. Kenyon et al (1983) obtained the distribution of G-C wavenumber along the long waves. This enables calculation of C_g and therefore wave energy using equations 4 and 5. Figure 5 Shows the distribution of G-C wavenumber along the long waves. Figure 6 shows the direction of the short waves orthogonals with respect to the long waves; in both figures 5 and 6, the parameters shown are calculated with respect to their initial value at the crest of the long wave with initial angle there taken arbitrarily to be 30°; the calculations are carried out for B between zero and .75. Figure 7 is a plan view of the propagation of the short waves on the long waves. Curving solid lines represent selected short wave crests. The initial angle of the wavenumber is 30° and B=0.75 here.

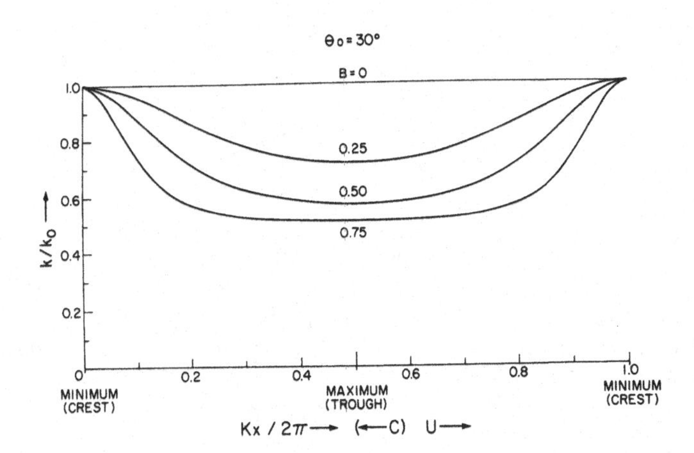

Figure 5. Distribution of short gravity-capillary wavenumber between two crests of the long waves. The initial incidence angle of the short waves is 30° at the crest, where the wavenumber is ko. B is a parameter related to long wave slope as described in the text (from Kenyon et al., 1983).

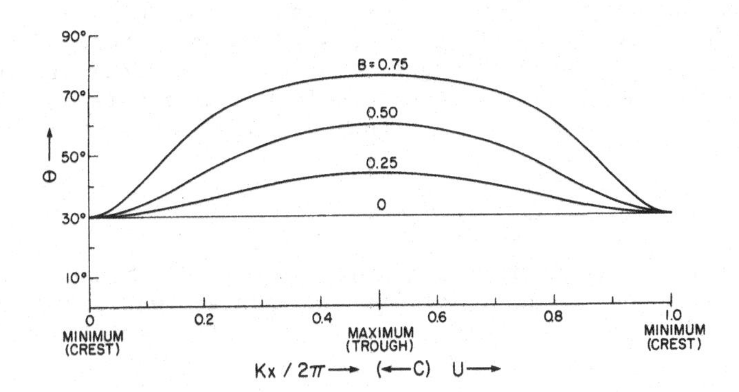

Figure 6. Distribution of short gravity-capillary incidence angle between two crests of the long waves. The initial incidence angle of the short waves is 30° at the crest. B is a parameter related to long wave slope, as described in the text (from Kenyon et al., 1983).

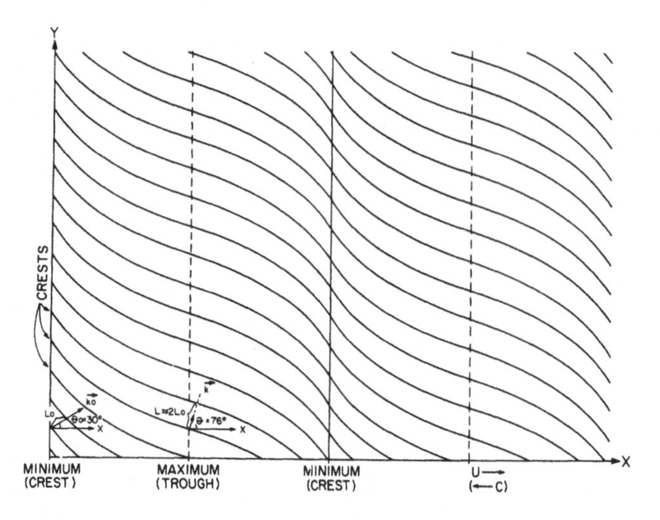

Figure 7. Plan view of the propagation of the short waves on the long waves. Curving solid lines represent selected short wave crests. The initial angle of the wavenumber of the short waves with the direction of the long waves - θ_o is 30° at the crest. B = .75 for this figure. (from Kenyon et al., 1983)

4. WIND STRESS

The increase in wind stress over water due to an increase in wave height has been recognized from measurements for some time (Hsu, 1974, Donelan, 1982, Geernaert, 1983). Different models have been proposed to help reduce the scatter in wind stress data collected over the ocean by parameterizing the effect of waves on wind stress. Most of these models assume that the waves propagate in the same general direction as the wind. The wind stress, τ_x, in a turbulent boundary layer (extending from a few millimeters to tens of meters above the ocean) is parameterized by

$$\tau_x = C_D(Z)\rho[\langle U(Z)\rangle - \langle U_o\rangle]^2 \qquad (7)$$

ρ is the air density, C_D is the drag coefficient, $\langle U(Z)\rangle$ and $\langle U_o\rangle$ are the mean wind speed at height Z and the mean surface water speed respectively. The drag coefficient during neutral stratification is presented as

$$C_D(Z) = \left(\frac{k}{\ln Z/Z_o}\right)^2 \qquad (8)$$

This is modified somewhat for non-neutral conditions. k~.4 is the von Karman constant, and Z_o is called the surface roughness length. The effect of waves on wind stress is usually incorporated via the roughness length. Hsu (1974) parameterized Z_o as proportional to the slope of the dominant wind waves. Geernaert (1983) calculated C_D as a function of wind speed for various dominant wave slopes using Hsu's parameterization and showed that an 80% increase in wave slope in a region of enhanced wave energy can increase C_D by 20% for a 15 m/s wind at Z=10m. In the adjacent low wave energy region, the wind stress is similarly reduced, depending on the reduced wave energy there. Donelan (1982) proposed that C_D can be expressed as the sum of long wave - C_{DL} and short wave - C_{DS} coefficients; $C_D = C_{DL} + C_{DS}$; C_{DL} is proportional to $[U(10m) - .83\ Cp/\cos\theta]\ |U(10) - .83\ cp/\cos\theta|$ where Cp is the phase velocity of relatively long waves and θ is the angle between the wave direction and the wind direction. Donelan's (1982) model estimates wind stresses that can be much larger than other models, due to long waves propagating away from or into the wind. There is no consensus on what is the most appropriate model, and it seems that more work needs to be done, theoretically and experimentally, to find out the effect of surface waves on the wind stress. It is clear, however, that the effect can be significant.

5. FEEDBACK MECHANISMS

From the discussion above, it is clear that circulation features affect the surface wave energy distribution and that this distribution, in turn, affects the wind stress, which is an important factor in forcing the circulation. Since this is good material for feedback, I will try to illustrate with an example. Horizontal velocity shear fronts act as a low pass filter for waves (Sheres et al., 1985). They block the short and relatively slow wind waves by total reflection or breaking, depending on the relative directions of the shear and the wind waves. This can be observed by the abrupt change in surface roughness, with the windward side of the shear much rougher than the lee side (Isaacs, 1948; Valenzuela et al., 1983), at such shears. As a result the windward side of the shear has a larger wind stress and forcing than the lee side. Depending on the wind direction relative to the shear, the wind stress can either enhance or reduce the shear flow, thus producing positive or negative feedback.

6. CONCLUSIONS

The results presented in figures 5-7 show the modulation of short G-C waves by much longer waves. Both wavelength and direction as well as the energy are modulated by the long waves. These effects are important for remote sensing, particularly if based on Bragg (resonance dependent on wavelength and direction) type scattering such as SAR. The calculation used here is idealized in order to isolate a particular factor in the C-G wave modulation; it remains to be investigated whether and under what conditions the above modulation mechanism is dominant. What is clear is that modulation of short G-C waves on scales of a few km can have their origin in mesoscale circulation and bathymetry features with much larger spatial scales. These modulations are not only relevant to remote sensing, but may also be important to forcing of the ocean due to interaction and perhaps feedback with wind stress.

7. ACKNOWLEDGMENTS

I thank Kenneth Reid and Richard McGill for the numerical work, Kern Kenyon and Jenifer Clark for the data, Gaspar Valenzuela for helpful discussions, Kathleen Funk for typing the manuscript, and the National Research Council and the Naval Research Laboratory for their support.

8. REFERENCES

Christoffersen, J. B., 1982: Current depth refraction of dissipative water waves. Series Paper No. 30, Institute of Hydrodynamics and Hydraulic Engineering, Technical University of Denmark.

Dobson, E., and D. Irvine, 1983: Investigation of Gulf Stream ring detection with spaceborne altimeter using mean sea height, wave height and radar cross section. Proceedings of the International Geoscience and Remote Sensing Symposium (IGARSS '83) held in San Francisco, on 31 August-2 September, 1, pp. TA-3 (6.1-6.7).

Donelan, M. A., 1982: The dependence of the aerodynamic drag coefficient on wave parameters. In the Procedings of: First International Conference on Meteorology and Air-Sea Interaction of the Coastal Zone, Am. Meteorology Soc., Boston, 381-387.

Geernaert, G. L., 1983: Variation of the drag coefficient and its dependence on sea state. Ph.d dissertation, University of Washington, Seattle, 203 pp.

Hollinger, J. P., (1971): Remote passive microwave sensing of the ocean surface. Proceedings of the seventh International Symposium on remote sensing of the environment, 3, Ann Arbor, Michigan, 1807-1817.

Hsu, S. A., 1974: A dynamic roughness equation and its application to wind stress determination at the air-sea interface. J. of Phys. Oceanography, 4, 116-120.

Isaacs, J. D., 1948: Discussion of "refraction of surface waves by currents" by J. W. Johnson. EOS Trans. AGU, **29**, 739-742.

Kenyon, K. E., 1971: Wave refraction in ocean currents. Deep-Sea Research, **18**, 1023-1034

Kenyon, K. E., D. Sheres and R. L. Bernstein, 1983: Short waves on long waves. J. of Geophysical Research, **83**, 7589-7596.

Mathiessen, M., 1987: Wave refraction by a current whirl. J. of Geophysical Research, **92**, No. C4, 3905-3912.

Scully-Power, P., J. Hughes and W. T. Aldinger, 1985: Navy Oceanographer Shuttle Observations. STS 41-G, Quicklook Report, Navy Underwater Systems Center Technical Document 7379.

Sheres, D., K. E. Kenyon, R. L. Bernstein, and R. C. Beardsley, 1985: Large horizontal surface velocity shears in the ocean obtained from images of refracting swell and in-situ moored current data. J. of Physical Oceanography, **12**, 200-207.

Sheres, D., D. T. Chen, and G. R. Valenzuela, 1987: Remote sensing of wave patterns with oceanographic implications. Int. Journal of Remote Sensing, **8**, No. 11, 1629-1640.

Sheres, D. and K. E. Kenyon, 1988: Swell refraction by the Pt. Conception, California eddy. submitted, Int. J. of Remote Sensing.

Teague, W. J., 1974: Refraction of surface gravity waves in an eddy. Master's thesis, University of Miami, Miami, Florida.

Valenzuela, G. R., D. T. Chen, W. D. Garrett, and J. A. C. Kaiser, 1983a: Shallow water bottom topography from radar imagery. Nature, **303**, 687-689.

Valenzuela, G. R., W. J. Plant, D. Schuler, D. T. Chen, and W. C. Keller, 1985: Microwave probing of shallow water bottom topography in the Nantucket Shoals. Journal of Geophysical Research, **90**, 4931-4942.

THE MAPPING OF UNDERWATER BOTTOM TOPOGRAPHY WITH SLAR

J. Vogelzang and D. Spitzer
Rijkswaterstaat, Tidal Waters Division
P.O. Box 20904
2500 EX The Hague
The Netherlands.

W.A. van Gein
Hydrographic Service Royal Netherlands Navy
Badhuisweg 171
2597 JN The Hague
The Netherlands

G.P. de Loor
Physics and Electronics Laboratory TNO
P.O. Box 96864
2509 JG The Hague
The Netherlands

H.C. Peters
Rijkswaterstaat, North Sea Directorate
P.O. Box 5807
2280 HV Rijswijk
The Netherlands

H. Pouwels
National Aerospace Laboratory
P.O. Box 90502
1006 BM Amsterdam
The Netherlands

G.J. Wensink
Delft Hydraulics
P.O. Box 152
8300 AD Emmeloord
The Netherlands

ABSTRACT. Recently an experiment has been performed to study the usability of mapping bottom topography with imaging radar for cartographic means. The experimental results can be compared with current theories of the imaging mechanism. The experiment was carried out in an area off the Dutch coast where the bottom topography is dominated by sand

G. J. Komen and W. A. Oost (eds.), Radar Scattering from Modulated Wind Waves, 231–242.

waves with a crest-to-crest distance of typically 300 m. These are small scale features, so advection might be important for the imaging mechanism. Therefore the theory of Alpers and Hennings is reconsidered. The action balance equation is solved numerically without any further approximation for a simple variation in the current velocity. It is shown that advection is important for small scale features like sand waves, especially at L-band. It is also shown that the source term in the action balance equation employed by Alpers and Hennings gives almost the same results as the one used by Shuchman, Lyzenga and Meadows.

1. INTRODUCTION

It is known for some time now that under suitable conditions (moderate wind and strong tidal current) the topography of the sea bottom can be made visible with imaging radars (Side Looking Airborne Radar, SLAR and Synthetic Aperture Radar, SAR). Since its discovery in 1969 by de Loor (de Loor and Brunsveld van Hulten, 1978 ; de Loor, 1981) this phenomenon has received much attention, especially after the SEASAT mission in 1978.

At the moment the mapping of sea bottom topography is studied as a part of the Dutch National Remote Sensing Program (NRSP). Recently an experiment has been performed to establish the usability of radar imagery of the sea for cartographic means. This experiment was set up in such a way that it is possible to check current theories of the imaging mechanism.

Several theories have been proposed to describe the imaging mechanism (Alpers and Hennings, 1984 ; Shuchman, Lyzenga and Meadows, 1985 ; van Gastel, 1987). The general idea is that the process of mapping sea bottom topography with imaging radar consists of three steps

1. the interaction between the tidal flow and the bottom topography produces variations in the current velocity at the sea surface. This is usually described with the continuity equations.
2. the variations in the surface current velocity give rise to modulations in the wind generated spectrum of water waves as described by the action balance equation.
3. the modulations in the wave spectrum cause variations in the radar backscatter.

Assuming first order Bragg-scattering to be the dominant mechanism, the backscattered intensity is proportional to the wave height spectrum at the Bragg wave number k_B given by

$$k_B = 2\ k \sin(\theta) \tag{1}$$

where k is the wave number of the incident radar radiation and θ the angle of incidence.

The simplest theory is the one by Alpers and Hennings (AH, 1984). They assume that variations in the surface current velocity lead to only small deviations of the wave spectrum from equilibrium. In a first order perturbation scheme and neglecting advection they arrive at an analytical solution of the action balance equation. Shuchman, Lyzenga and

Meadows (SLM, 1985) solve the action balance equation numerically using the source term given by Hughes (1978). Van Gastel includes the effects of wind-input, dissipation and non-linear wave-wave interactions in a rigorous way, finding large modulations in the radar backscatter at X-band.

Section 2 contains a description of the experiment. It was performed in an area dominated by sand waves with a crest-to-crest distance of typically 300 m. Since these sand waves are small scale features, advection might be important for the imaging mechanism. The action balance equation is discussed in section 3. This equation is solved numerically without any further approximation for the source term given by AH, to check the range of validity of their results. The results for a simple variation in the current velocity are presented in section 4. It will be shown that advection is important for small scale features like sand waves. It is also shown that the source term in the action balance equation employed by AH leads to almost the same results as the one used by SLM.

2. THE EXPERIMENT

The experiment has been performed on January 19, 1988, with the Dutch digital SLAR operated by the National Aerospace Laboratory (NLR). With this system and the image processing facilities at NLR both geometrically and radiometrically correct images can be obtained. Figures 1a-1d show examples of such images, taken on April 26, 1985. Figure 1a shows the Hooge Platen, a sand bank in the Westerschelde. Figure 1b shows the Westerschelde between Vlissingen (above) and Breskens (below). Beside ships (large white spots) and buoys (small white dots) also bottom topography features are visible. Figures 1c and 1d are taken near Westkapelle (to the right). The light spots near the middle of the images are caused by breaking waves above the Rassen, an area with a depth of 1 m or less. Between the Rassen and the coast the Oostgat, a narrow and deep ship route is clearly visible.

It would be very difficult to compare theoretical predictions with images such as figures 1a-1d, due to the complex bottom topography and streaming patterns. One would like an area with a simple, quasi one-dimensional topography to make model calculations needed to check the current theories feasible. Moreover, most existing theories are one-dimensional.

Therefore the experiment has been performed in an area 20 km NW to the Noordwijk Platform (MPN). The bottom topography of this area is dominated by sand waves with a height of 2-6 m and a crest-to-crest distance of typically 300 m at a depth of about 23 m. The testing area consists of two mutually perpendicular rectangles with common centre, each measuring 5 x 10 km. The direction of the dominating current velocity is parallel to the coast line and perpendicular to the sand wave crests. Two flights were made to study the effect of the current velocity direction : one at low tide and one at high tide, when the current velocity is maximal. During each flight both rectangles were recorded from both flight directions parrallel to the longer sides of

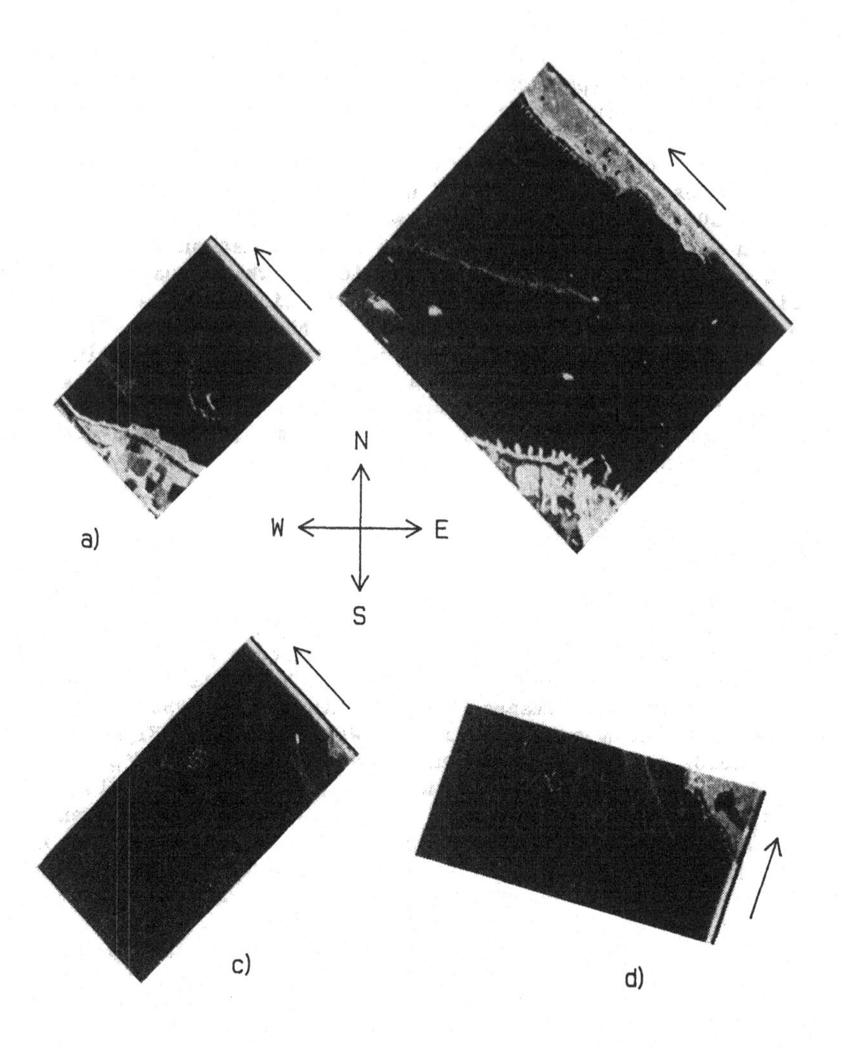

Figure 1. Dutch digital SLAR images of the Westerschelde. The arrows indicate theflight direction. a) Hooge Platen, b) Westerschelde between Vlissingen and Breskens, c) and d) Rassen near Westkapelle.

the rectangles to study the influence of the flight direction. Simultaneously the current velocity profile was measured from two ships operated by Rijkswaterstaat, one located above a crest and the other above a trough. By extrapolation the current velocity at the surface can be obtained from the profile. Note that one measurement of the current velocity at the surface suffices to determine the current velocity in the whole area if the interaction between bottom topography and tidal flow can be described with the continuity equations.

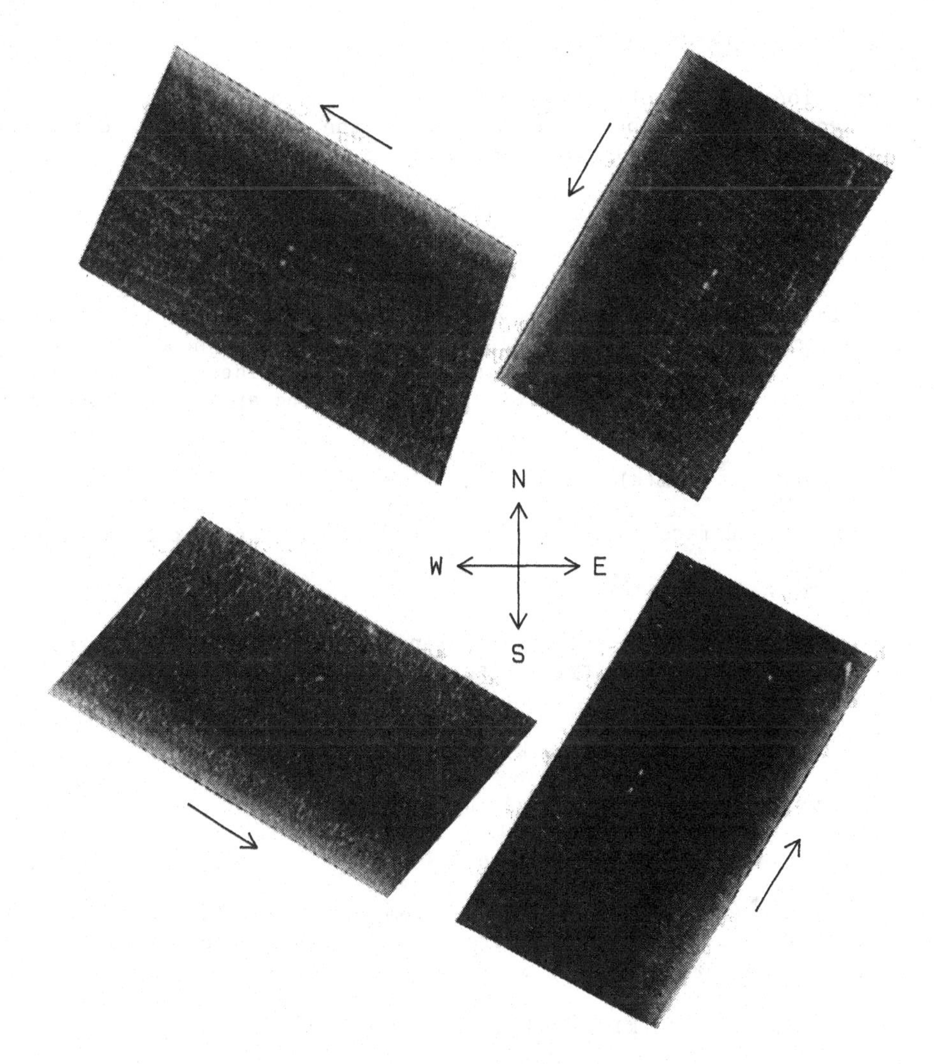

Figure 2. Dutch digital SLAR images of sand waves 20 km NW to MPN. The arrows indicate the flight direction.

Figure 2 shows the images obtained at high tide. The current velocity at the surface exceeded 0.5 m/s, the wind speed was 8 m/s. This rather high wind speed was necessary to receive enough backscatter from the sea surface, due to the low sensitivity of the radar system. The comparison of these images with digitized bathymetric maps is in progress. In a later stage theoretical predictions will be compared with the images by model calculations.

3. THE ACTION BALANCE EQUATION

The evolution of the action spectrum A is described by the action balance equation (Willebrand, 1975 ; Hasselmann, 1960). In one dimension and for steady state this equation reads :

$$\frac{\partial\Omega(k,x)}{\partial k}\frac{\partial A(k,x)}{\partial x} - \frac{\partial\Omega(k,x)}{\partial x}\frac{\partial A(k,x)}{\partial k} = S(k,x) \quad , \tag{2}$$

where x is the position, k the wave number and S the source term describing the combined effect of wind input, dissipation and nonlinear wave-wave interactions. The first term on the left hand side of (2) is the advection term, the second the refraction term. The apparent frequency Ω is given by

$$\Omega(k,x) = \omega(k) + k\,U(x) \quad , \tag{3}$$

where U is the surface current velocity and ω the intrinsic frequency

$$\omega(k) = (g\,k + \tau\,k^3)^{1/2} \quad , \tag{4}$$

g being the gravitational acceleration and τ the ratio of the surface tension and the water density. The action spectrum is related to the energy spectrum E by

$$A(k,x) = E(k,x)\,/\,\omega(k) \quad . \tag{5}$$

The source function S is given by AH as

$$S(k,x) = -\,\mu\,[A(k,x) - A_o(k)] \quad , \tag{6}$$

where A_o is the equilibrium action spectrum and μ the relaxation rate or wave growth rate parameter, considered as an adjustable parameter in their model. The action balance equation (2) is solved by assuming

$$A(k,x) = A_o(k) + \delta A(k,x) \quad , \tag{7a}$$

$$U(x) = U_o + \delta U(x) \quad , \tag{7b}$$

where δA and δU are small compared to A_o and U_o respectively. Neglecting advection and second order terms in δA and δU they arrive at an analytical solution, valid if

$$(c_g + U_o)\,K \ll \mu \quad , \tag{8}$$

with c_g the group velocity and K the wave number of the bottom topography. This implies that the Alpers and Hennings model might fail for small scale features.

The source function as given by SLM is

$$S(k,x) = \mu(k)\, A(k,x)\, [1 - A(k,x)/A_o(k)] \quad . \tag{9}$$

Here the wave growth rate μ depends on the wave number as well as the wind speed. The wave growth rate is not very well known. Available measurements of it have been reviewed by Hughes (1978) and fitted by an expression which for parallel wind and wave direction reads

$$\mu(k) = \omega(k)\, u_*/c\, [0.01 + 0.016\, u_*/c]$$

$$\{1 - \exp[-8.9\, (u_*/c - 0.03)^{1/2}]\} \tag{10}$$

where c is the phase speed of the waves and u_* the friction velocity. Figure 3 shows μ as a function of U_{10}, the wind speed at 10 m anemometer height, for $k = 30\ m^{-1}$ and $k = 300\ m^{-1}$. The curves in figure 3 were obtained using the relations between U_{10} and u_* riven by Garratt (1977) and Inoue (1967). In the calculations presented in the next chapter μ is considered as a constant; the wave number and wind speed dependence (10) is not taken into account.

Note that substitution of (7a) in (9) and neglecting quadratic terms in δA yields equation (6). One therefore expects the two models to be similar if the modulation δA is small.

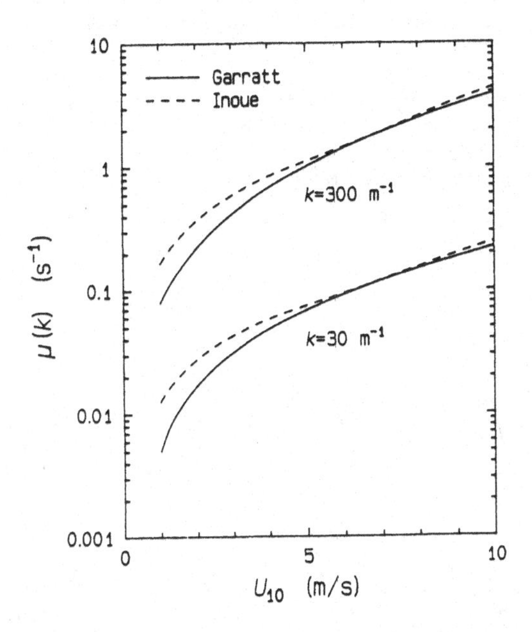

Figure 3. The parametrization of μ by Hughes as a function of U_{10} with the relations between U_{10} and u_* given by Garratt and Inoue.

The action balance equation (2) is solved following the method of characteristics employed by SLM and Hughes. The surface current velocity had the following form :

$$U(x) = U_o + U_m \cos(2\pi x/b) \quad , \tag{11}$$

with b the length of the current velocity variation. In the calculations presented in the next section we took U_o = 1 m/s and U_m = 0.1 m/s.

4. RESULTS AND DISCUSSION

Figure 4 shows the relative variations ΔU and ΔA,

$$\Delta U = [U(x) - U(0)]/U(0) \quad , \tag{12a}$$

$$\Delta A = [A(k,x) - A(k,0)]/A(k,0) \quad , \tag{12b}$$

as a function of the position. The curves for ΔA were calculated with k = 30 m^{-1}, corresponding to the Bragg wave number for radar radiation with a wave length of 30 cm (L-band) at an angle of incidence of approximately 45°, and with k = 300 m^{-1}, corresponding to the Bragg wave number for 3 cm wavelength radar radiation (X-band) incident under the same angle. The values for μ were chosen from figure 3, using Garratts parametrization of the roughness length at a wind speed of 5 m/s. From figure 4 one can see that the modulation in the action spectrum ΔA

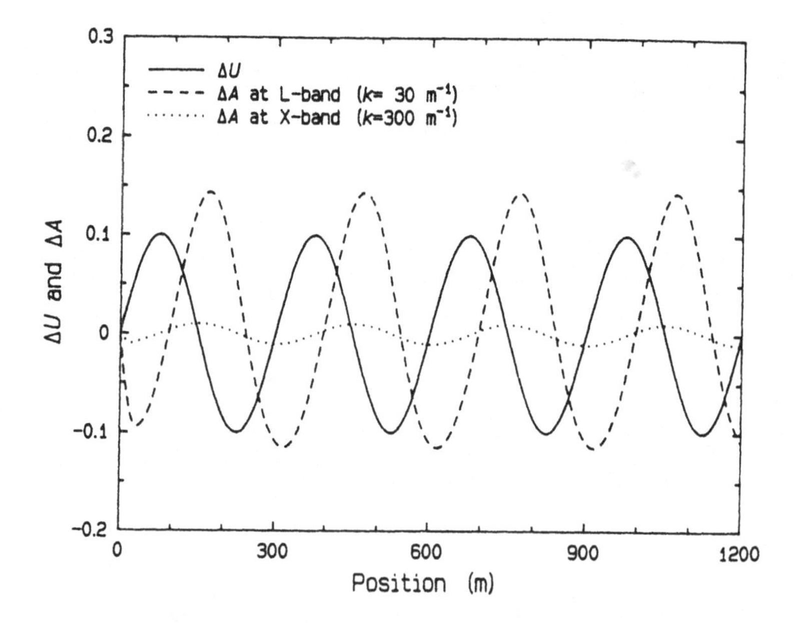

Figure 4. ΔA and ΔU as a function of the position.

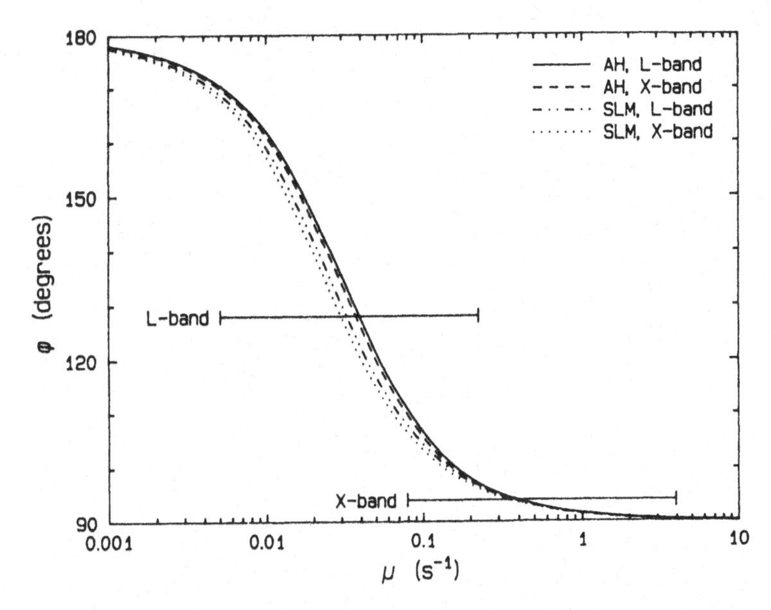

Figure 5a. The phase shift φ as a function of μ. The bars indicate the physical values of μ.

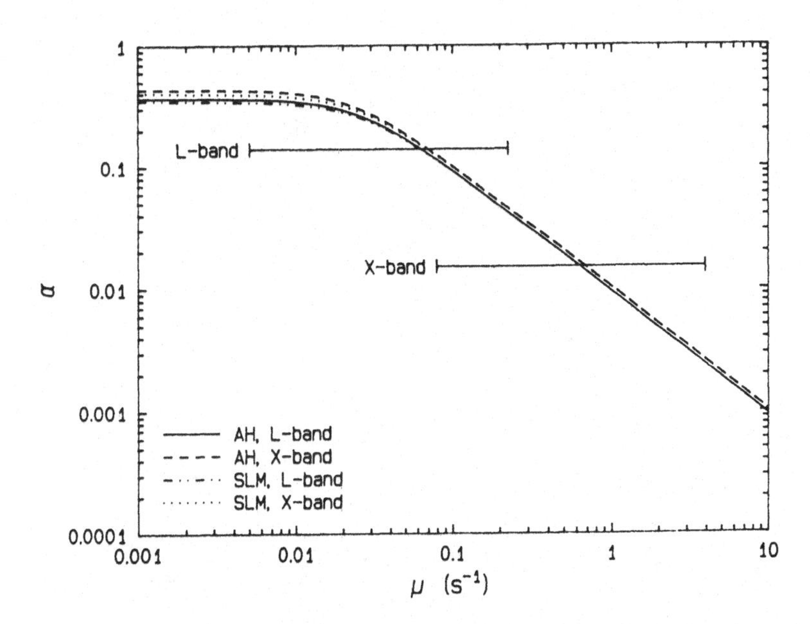

Figure 5b. The modulation depth α as a function of μ. The bars indicate the physical values of μ.

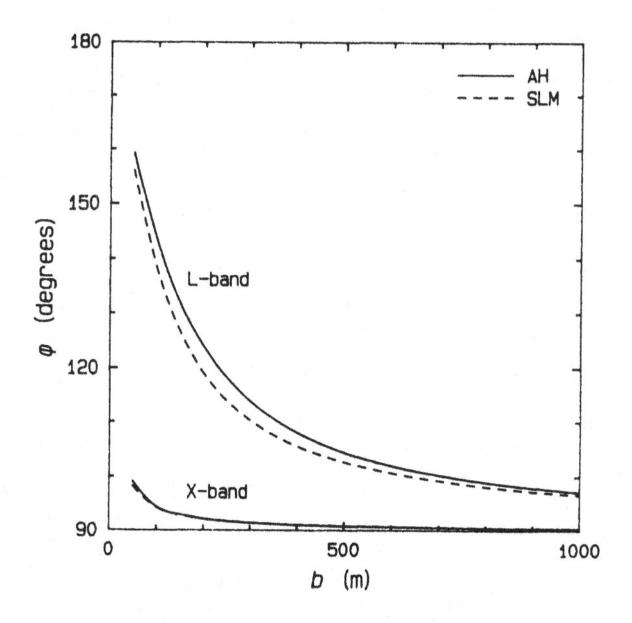

Figure 6a. The phase shift φ as a function of b.

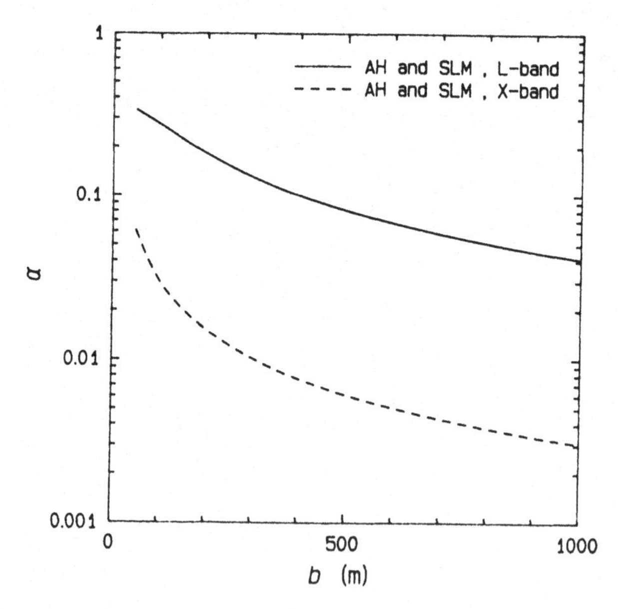

Figure 6b. The modulation depth α as a function of b.

follows the modulation in the current velocity ΔU. Therefore the modulation depth α is defined as half the difference of the extreme values of ΔA, and the phase shift φ as the difference of the positions of the maxima of ΔA and ΔU times $2\pi/b$. Assuming first order Bragg scattering to be the dominant scattering mechanism, the relative change in intensity equals ΔA.

Note that the modulation depth decreases with increasing relaxation rate, and therefore with increasing wind speed. This might explain the poor quality of the experimental results (figure 2) compared to older measurements in the same area with a more sensitive radar at lower wind speeds (de Loor and Brunsveld van Hulten, 1978 ; de Loor, 1981).

Figure 5a shows φ as a function of μ for b = 300 m. The bars indicate the range of physical values of μ for wind speeds between 0.5 and 10 m/s as calculated from (10). For X-band ($k = 300\ m^{-1}$) φ ranges from 90° to 110°, while for L-band ($k = 30\ m^{-1}$) φ ranges from 100° to 170°. This contradicts the result of AH that φ = 90°, the difference being due to advection. Figure 5b shows α as a function of μ, also for b = 300 m. Though the modulation at X-band is larger than at L-band for a fixed value of μ, the predicted modulation at X-band will be an order of magnitude smaller than at L-band due to the different physical values of μ indicated by the bars.

Figure 6a shows φ as a function of b at L-band ($k = 30\ m^{-1}$) and at X-band ($k = 300\ m^{-1}$). The values of μ were chosen from figure 3 using the parametrization of the roughness length given by Garratt at a wind speed of 5 m/s. For smaller scale features the phase angle deviates more and more from the value of 90° predicted by AH, notably at L-band. Figure 6b shows α as a function of b for the same values of μ and k as in figure 6a. The modulation at X-band is an order of magnitude smaller than at L-band. The source functions of AH and SLM give almost the same results.

AH obtain good agreement with SEASAT L-band images of large banks using $\mu = 0.025\ s^{-1}$ at a Bragg wave number of $18.5\ m^{-1}$. They find a modulation depth up to 0.4, consistent with our results. Figures 5 and 6 show that the source term of AH leads to almost the same results as the one of SLM if advection is taken into account.

5. CONCLUSIONS

From the model calculations presented above we concude that

- advection is important for features smaller than 1000 m, giving rise to a larger phase shift than predicted by Alpers and Hennings, notably at L-band.
- the source term of Alpers and Hennings leads to almost the same results as the one of Shuchman, Lyzenga and Meadows.
- the modulation at X-band is an order of magnitude smaller than at L-band if the parametrization of Hughes for the wave growth rate parameter is used.

It should be noted that the modulation at X-band is increased drastically by including scattering mechanisms other than Bragg-scattering (Holliday, St-Cyr and Woods, 1986) or taking non-linear wave-wave interactions into account (van Gastel, 1987).

ACKNOWLEDGEMENTS

This work has been supported by the Dutch National Remote Sensing Board (BCRS) as a part of the National Remote Sensing Program (NRSP). The authors would like to thank G.J. Komen for his interest and stimulating discussions.

REFERENCES

ALPERS, W. and HENNINGS, I., 1984, 'A theory of the imaging mechanism of underwater bottom topography by real and synthetic aperture radar', J. Geophys. Res. **89 C** 10529-10546.

GARRATT, J.R., 1977, 'Review of drag coefficients over oceans and continents', Monthly Weather Review **105** 915-929.

Van GASTEL, K., 1987, 'Imaging by X-band radar of subsurface features : a nonlinear phenomenon', J. Geophys. Res. **92 C** 11857-11865.

HASSELMANN, K., 1960, 'Grundgleichungen der Seegangsvorhersage', Schiffstechnik **7** 191-195.

HOLLIDAY, D., ST-CYR, G. and WOODS, N.E., 1986, 'A radar ocean imaging model for small to moderate incidence angles', Int. J. Remote Sensing **7** 1809-1834.

HUGHES, B.A., 1978, 'The effect of internal waves on surface wind waves. 2 Theoretical analysis', J. Geophys Res. **83 C** 455-465.

INOUE, T., 1967, 'On the growth of the spectrum of a wind generated sea according to a modified Miles-Phillips mechanism and its application to wave forecasting', Rep. TR-67-5, Geophys. Sci. Lab., New York Univ., New York.

De LOOR, G.P. and BRUNSVELD van HULTEN, H.W., 1978, 'Microwave measurements over the North Sea', Bound. Lay. Met. **13** 113-131.

De LOOR, G.P., 1981, 'The observation of tidal patterns, currents and bathymetry with SLAR imagery of the sea', IEEE Trans. Geosci. Electron. **OE-6** 124-129.

SHUCHMAN, R.A., LYZENGA, D.R. and MEADOWS, G.A., 1985, 'Synthetic aperture radar imaging of ocean-bottom topography via tidal-current interactions : theory and observations', Int. J. Remote Sensing **6** 1179-1200.

WILLEBRAND, J., 1975, 'Energy transport in a nonlinear and inhomogeneous random gravity wave field', J. Fluid Mech. **70** 113-126.

Session 5

General discussion

GENERAL DISCUSSION ON ELECTROMAGNETIC SCATTERING

Gaspar R. Valenzuela, Discussion Leader
Space Sensing Branch
Naval Research Laboratory
Washington, DC 20375-5000
U.S.A.

SUMMARY

The purpose of this general discussion on e.m. scattering after the formal scientific presentations at the workshop was to ascertain the present state of knowledge in the field, assess the quality of the predictions, and to arrive at some specific recommendations for future research. Frank and active discussions took place in regard to the status of e.m. scattering theories/models for the ocean surface (see Table I). During the formal presentations it was clear that the two-scale model (Wright, 1968; Bass and Fuks, 1979) when properly applied, (Thompson, 1988; Thompson et al., 1988) yields just as good agreement (within a couple of decibels) as the Kirchhoff approximation (Holliday et al., 1986,1987) when applied to the SARSEX data on SAR imagery (L and X band) for internal waves in the New York Bight. Both of these comparisons only deal with cross section effects, the distortion in the image intensity from wave motion and current variations affecting the phase of the e.m. backscattered fields is not included. To match the theoretical predictions with the measurements, the enhancement in slopes by the internal waves of both the long (greater than 1 m length) and intermediate (of the order of 1 m length) gravity waves should be included in the tilting of the Bragg resonant waves of wave number $2k_0 \sin\theta$, k_0 is the wavenumber of the e.m. radiation in free space and θ is the angle of incidence.

Of course, a number of other scattering mechanisms should still be included in scattering theories, such as wave breaking, wedge scattering, volume scattering, etc. In any case, the question on how to proceed to improve the available e.m. scattering theories and models to predict/interpret ocean field observations was brought up. The following recommendations were agreed upon:

1. Quantitative testing needed of e.m. scattering theories and models for the ocean in well instrumented wave tanks and ocean field towers.
2. Have a workshop on the mathematical aspects of e.m. scattering theories for the ocean.

G. J. Komen and W. A. Oost (eds.), Radar Scattering from Modulated Wind Waves, 245–247.

TABLE I - AVAILABILITY OF SCATTERING THEORIES/MODELS FOR THE OCEAN SURFACE (BACKSCATTER)

	TWO-SCALE MODEL (Bragg & Specular)	KIRCHHOFF APPROXIMATION (One Iteration for Surface Fields)	KIRCHHOFF APPROXIMATION (Two Iterations for Surface Fields)	INTEGRAL EQUATIONS (Stratton-Chu)	GREEN'S FUNCTION METHOD
Perfectly Conducting Surfaces	YES	YES	YES	YES (FORMALLY)	YES (FEYNMAN DIAGRAMS)
Dialectric Surfaces	YES	YES	NOT DEVELOPED	YES (FORMALLY)	NO
Polarization Effects	YES	NO	NOT DEVELOPED	NOT DEVELOPED	NOT DEVELOPED
Doppler Spectrum	YES	YES	NOT DEVELOPED	NOT DEVELOPED	NOT DEVELOPED
Range of Validity-Angle of Incidence (Degrees)	0-85	0-45	0-?	0-90	0-90

3. More cooperation required between national and international interdisciplinary groups to address the problems relating to ocean surface signatures in remote sensing.

Nonetheless, ultimately what is needed by the remote sensing community is the development of more rigorous and accurate e.m. scattering models for the ocean for all radar frequencies. These probably could evolve by systematic improvement of the available two-scale model and the Kirchhoff approximation, or from further simplification and reduction of the more exact Integral Equations and Green's Function Method. We conclude by cautioning, that even with an exact scattering theory, the quality of the predictions is no better than the characterization (in space and time) of the ocean surface.

REFERENCES

Bass, F.G. and I.M. Fuks, 1979: Wave Scattering from Statistically Rough Surfaces. Pergamon Press, 525 pp.

Holliday, D., G. St-Cyr and N.E. Woods, 1986: A radar ocean imaging model for small to moderate incidence angles. Int. J. Remote Sensing, **7**, 1809-1834.

Holliday, D., G. St-Cyr and N.E. Woods, 1987: Comparison of a new radar ocean imaging model with SARSEX internal wave image data. Int. J. Remote Sensing, **8**, 1423-1430.

Thompson, D.R., 1988: Calculation of radar backscatter modulations from internal waves. J. Geophys. Res., **93**, in press.

Thompson, D.R., B.L. Gotwols and R.E. Sterner, 1988: A comparison of measured surface wave spectral modulations with predictions from a wave-current interaction model. J. Geophys. Res., **93**, in press.

Wright, J.W., 1968: A new model for sea clutter. IEEE Trans., AP-16, 217-223.

GENERAL DISCUSSION ON THE ENERGY BALANCE IN SHORT WIND WAVES

William J. Plant
Naval Research Laboratory
Washington, DC 20375
USA

THEORETICAL CONSIDERATIONS

The generally-accepted approach to the problem of the energy balance in short wind-generated waves is to use the action conservation equation. Most uncertainty enters this procedure in the selection of forms for the source terms on the right hand side of this equation although the equation can be solved to second-order in long wave slope in which case questions of long wave/short wave interactions enter the problem. The three areas encompassing all these source terms are wind input, wave/wave interactions, and dissipation.

Wind input. The group generally agreed that this is the best understood of the source terms since much data on wave growth rates is in hand. Some concern was expressed that some energy could be extracted from the wind in forms not amenable to the standard methods of measuring growth rates and thus that some wind input could be neglected. It was agreed, however, that input terms much different from measured growth rates times spectral density would be very suspect.

Wave/wave interactions. The general approach when wavelengths do not differ too much is to use the standard, but very complex, perturbation theory result which yields energy transfer due to quartet interactions for gravity waves and triplet interactions for gravity-capillary waves. Concern was expressed both over the complexity of this approach and the fact that wind shear effects are generally omitted from these calculations. The group agreed that this theory was fairly well in hand, however.

When modulation of the short waves by much longer waves is included in the calculation, knowledge of modulation transfer functions becomes important. At present, these functions are well known empirically for a very select region of the relevant parameter space and a theoretical explanation for their magnitude and phase is still not in hand. Thus, if such interactions are important in the overall

G. J. Komen and W. A. Oost (eds.), Radar Scattering from Modulated Wind Waves, 249–251.

energy balance, they can only be included empirically at present.

Dissipation. The group agreed that this is the biggest problem in determining the overall energy balance in short wind-generated waves. Present theories account for this using dissipation functions which are proportional to total wave energy to an arbitrary power times spectral density, to spectral density squared, and to spectral density cubed. Most workshop participants agreed that some such nonlinear form was necessary to explain the growth and equilibrium of short wind waves. Each of these forms has its adherents, however, and each seems to work to some extent in selected regions of the spectrum. Remarks concerning the difficulty of inferring the form of this function from the overal energy balance due to experimental uncertainties were met with the observation that the new, area-extensive optical techniques presently being developed may make such studies feasible. The suggestion was made that some indication of the form of the dissipation function could be obtained by creating mechanical turbulence in the water and observing its effect on the waves. It was agreed, however, that this was not exactly the same dissipation as would occur in the presence of wind-induced turbulence.

Experimental Considerations

Historically, point measurements of waves have yielded frequency spectra from which most information about waves have been obtained. Such techniques include surface piercing wave gauges and laser slope gauges. Most participants agreed that the information which could be obtained from such techniques was rather limited. Doppler shifts caused by long waves must be removed from the measurements and it is, in general, difficult to obtain directional information from the techniques. Adherents pointed out the ease of making such measurements, especially in the field, which makes them quite benefical for calibration purposes and obtaining gross wave properties. Most workshop participants agreed that converting from frequency to wavenumber spectra using a simple dispersion relationship was a questionable procedure and had only been attempted in the past for want of better techniques.

More exciting to the group were the area-extensive measurement techniques currently in use or under development. These techniques include CCD video techniques using graded light sources below the surface and stereo photography. Such techniques have the capability of taking two dimensional images of the short waves up to thirty times a second, a rate which allows the complete short wave spectrum as a function of both wavenumber and frequency to be obtained. This provides nearly complete information about the wave field. Such techniques are still difficult to apply in the field, however, and are expensive and time-consuming to analyze once the data have been collected. The question of the effect of long wave tilt on the measured short wave spectra was raised. The general feeling was that this was not a serious problem and was one which could be corrected in the processing anyway. Thus the proper short wave spectrum was declared to be that one for which long wave tilt had been taken out. It was pointed out that this was essentially a two-scale procedure since some determination must be made of the spatial scale to be called long and for which tilt would be corrected. Thus it appears difficult to define an exact short wave spectrum without invoking a two-scale approximation.

Finally, the question of how to handle a disconnected surface was raised. At some wind speed, spray, foam, bubbles, and wave breaking become so pronounced that a surface cannot even be defined. Instead of a step function discontinuity in density, the density somehow slowly increases from that of air to that of water over a substantial vertical distance. Thus the concepts of surface displacement and wave spectra become inapplicable. All meeting participants agreed that this question was an important one which at present could not be adequately addressed. A viable description of a disconnected air/sea interface must await future research.

General discussion on Perturbation of the gravity-capillary wave spectrum by current variations

Discussion leader
Klaartje van Gastel
Mathematisch Instituut RUU
Postbus 80 010 3508 TA Utrecht
the Netherlands

During this workshop several models were presented to explain the modulation of short wind waves by non-uniform currents. For details on these models, see the contributions by Alpers, Thompson, Thomas & Edwards and Gastel. All models can be seen as limiting cases of the complete energy balance. They are obtained by neglecting some terms and making assumptions on the functional description of others. Each model can be extended by including the effect of large scale waves.

In the discussion it was attempted to determine which of the effects, dominating in the separate models, are necessary to explain the data. This was done by stating for each model it's main predictions and the theoretical estimate of the range of validity, i.e. the conditions in which it can be tested.

The outcome of the discussion was that, as they stand, each model can predict striking agreement with some of the data, but for all cases exist in which they are off. Therefore, a recommendation is that a unifying theory be developed, encompassing all existing models. This could be achieved by starting anew from the complete energy balance. To guarantee a thorough understanding of the waves this unifying theory should also be able to predict measured equilibrium spectra.

G. J. Komen and W. A. Oost (eds.), Radar Scattering from Modulated Wind Waves, 253–254.

A further recommendation is that experiments be done to illuminate the physics contained in the models. The parameters present in the models would have to be chosen carefully. These are: wavenumber range, wind strength, horizontal current and current shear, length of the modulation, angle between wind and current, and sea state. In the discussion vertical current shear and surface tension were also recognised to be of importance. The experiments should give as much information as possible on the development of the energy spectrum before and during it's modulation. This development might be characterized by 1. depth of the modulation of the energy density, 2. phase of this modulation, 3. shift of the mean (taken over a modulation cycle) of the energy density compared to the background and 4. sensitivity to wavenumber.

The general attitude was that we are quite advanced in our understanding of the modulation at L-band, but the shorter waves still present many challenges.

Subject index

Author index

List of participants

Prof. W. Alpers
Universität Bremen
Postfach 330440
2100 BREMEN 33
FRG

Dr. C.J. Calkoen
Delft Hydraulics
Postbus 152
8300 AD EMMELOORD
The Netherlands

Mr. Dierking
Universität Bremen
Postfach 330440
2100 BREMEN 33
FRG

Dr. H. Dolezalek
US Office of Naval Research
ARLINGTON, Va. 22217-5000
USA

Dr. K. van Gastel
Mathematisch Instituut
Postbus 80010
3508 TA UTRECHT
The Netherlands

Dr. G.L. Geernaert
US Naval Research Laboratory
Code 8314
WASHINGTON D.C. 20375
USA

Dr. David A. Greenberg
Bedford Institute of Oceanography
Coastal Oceanography Division
P.O. Box 1006
DARTMOUTH, Nova Scotia B2Y 4A2
Canada

Mr. D. van Halsema
Fysisch en Electronisch Lab. TNO
Postbus 96864
2509 JG DEN HAAG
The Netherlands

Dr. Ingo Hennings
GKSS Forschungszentrum
Geesthacht GmbH
Institut für Physik
Postfach 1100
2054 GEESTHACHT
FRG

Dr. S.J. Hogan
Mathematical Institute
24-29 St. Giles
OXFORD
England

Dr. D. Holliday
R&D Associates
P.O. Box 9695
MARINA DEL REY, CA 90295
USA

Dr. B. Jähne
Scripps Institution of
Oceanography
University of California at
San Diego
LA JOLLA, CA 92093-0212
USA

Dr. P.L.C. Jeynes
Oxford Computer Services Ltd
Flat 14
21 Lexham Gardens
LONDON W8 5JJ
United Kingdom

Prof. K.B. Katsaros
Dep. of Atmospheric Sciences
(AK-40)
University of Washington
SEATTLE, WA 98195
USA

Dr. G.J. Komen
KNMI
Postbus 201
3730 AE DE BILT
The Netherlands

Dr.Ir. G.P. de Loor
Fysisch en Electronisch Lab. TNO
Postbus 96864
2509 JG DEN HAAG
The Netherlands

Dr. W.A. Oost
KNMI
Postbus 201
3730 AE DE BILT
The Netherlands

Ir. H.C. Peters
Rijkswaterstaat
North Sea Directorate
Postbus 5807
2280 HV RIJSWIJK
The Netherlands

Dr. W.J. Plant
US Naval Research Laboratory
Code 8314
WASHINGTON, DC 20375
USA

Dr. W. Rosenthal
GKSS-Forschungszentrum
Geesthacht GmbH
Max Plank Strasse
Postfach 1100
D-2054 GEESTHACHT
FRG

Dr. O.H. Shemdin
Ocean Research and Engineering
255S. Marengo Ave.
PASADENA, CA 91101
USA

Dr. D. Sheres
Naval Research Laboratory
Code 8312
WASHINGTON D.C. 20375
USA

Cdr. J.P. Simpson
Office of Naval Research, London
Box 39
FPO
NEW YORK 09510
USA

Mr. P. Snoey
TU Delft
Fac. of Telecommunication and
Teleobservation Technology
Postbus 5031
2600 GA DELFT
The Netherlands

Dr. M.A. Srokosz
British National Space Center
Building R16
Royal Aircraft Establishment
Farnborough
HAMPSHIRE GU14 6TD
United Kingdom

Dr. Siegfried Stolte
Forschungsanstalt der Bundeswehr
für Wasserschall- und Geophysik
Klausdorfer Weg 2-24
2300 KIEL 14
FRG

Dr. J.O. Thomas
Oxford Computer Services Ltd.
52 St. Giles'
OXFORD OX1 3LU
United Kingdom

Dr. D.R. Thompson
Applied Physics Lab.
The Johns Hopkins University
LAUREL, MD 20707
USA

Dr. G.R. Valenzuela
code 8312
Naval Research Lab.
WASHINGTON, DC 20375-5000
USA

Mr. J. Vogelzang
Rijkswaterstaat
Dienst Getijde Wateren
Postbus 20904
2500 EX 's-GRAVENHAGE
The Netherlands

9 780792 301462